QUANTUM THEORY OF CHEMICAL REACTIONS

QUANTUM THEORY
OF
CHEMICAL REACTIONS

I. Collision Theory, Reaction Path, Static Indices

Edited by

RAYMOND DAUDEL

CNRS, Centre de Mécanique Ondulatoire Appliquée, Paris, France

and

ALBERTE PULLMAN

CNRS, Institut de Biologie Physico-Chimique, Paris

LIONEL SALEM

CNRS, Laboratoire de Chimie Théorique, Université de Paris Sud, Orsay

ALAIN VEILLARD

CNRS, Université Louis Pasteur, Strasbourg

D. REIDEL PUBLISHING COMPANY

DORDRECHT : HOLLAND / BOSTON : U.S.A.

LONDON : ENGLAND

Library of Congress Cataloging in Publication Data

Main entry under title:

Quantum theory of chemical reactions.

    Includes indexes.
    CONTENTS: v. 1   Collision theory, reaction path, static indices.
    1. Quantum chemistry – Addresses, essays, lectures.
I. Daudel, Raymond.
QD462.5.Q38   541'.28   79–22914
ISBN-13: 978-94-009-9518-5   e-ISBN-13: 978-94-009-9516-1
DOI: 10.1007/ 978-94-009-9516-1

Published by D. Reidel Publishing Company,
P.O. Box 17, Dordrecht, Holland.

Sold and distributed in the U.S.A., Canada and Mexico
by D. Reidel Publishing Company, Inc.
Lincoln Building, 160 Old Derby Street, Hingham,
Mass. 02043, U.S.A.

TABLE OF CONTENTS

PREFACE

This treatise is devoted to an analysis of the present state of the quantum theory of chemical reactions.

It will be divided into three volumes and will contain the contributions to an international seminar organized by the editors.

The first one, is concerned with the fundamental problems which occur when studying a gas phase reaction or a reaction for which the solvent effect is not taken into account.

The two first papers show how the collision theory can be used to predict the behaviour of interacting small molecules.

For large molecules the complete calculations are not possible. We can only estimate the reaction path by calculating important areas of the potential surfaces. Four papers are concerned with this important process. Furthermore, in one of these, the electronic reorganization which occurs along the reaction path is carefully analyzed.

Two papers are devoted to the discussion of general rules as aromaticity rules, symmetry rules.

The last two papers are concerned with the electrostatic molecular potential method which is the modern way of using static indices to establish relations between structure and chemical reactivity.

Volume II will be devoted to a detailed analysis of the role of the solvent and volume III will present important applications as reaction mechanisms, photochemistry, catalysis, biochemical reactions and drug design.

SOME RECENT DEVELOPMENTS IN THE MOLECULAR TREATMENT OF ATOM-ATOM COLLISIONS.

V. Sidis

Centre de Mécanique Ondulatoire Appliquée - 23, rue du Maroc, 75019 Paris - France
and Laboratoire des Collisions Atomiques, Université Paris XI, Bât 351, 91405 Orsay - France

INTRODUCTION

Since a few years, we have been applying (at the C.M.O.A.[1] in Paris and at the L.C.A.O.[2] in Orsay) Molecular Physics Methods to the study of atom-atom or ion-atom collisions at low and moderate energies (from few eV to few keV). The procedure is somewhat similar to that used in the study of Electronic Spectra of Diatomic Molecules. Indeed, in the velocity range under consideration, electrons move generally much faster than the nuclei which suggests to find approximate electronic wavefunctions for the clamped nuclei problem. These functions, in turn, serve as an expansion basis for expressing the total wavefunction that describes both the electrons and the nuclei of the system. The study of an atom-atom collision problem then requires the resolution of a more or less large set of coupled equations. These equations determine the nuclear wavefunctions whose assymptotic behavior yield the scattering amplitudes and consequently the cross sections.

Actually, since a collision problem differs from a bound diatomic molecule in the spatial extension of the nuclear motion and in the velocity of the nuclei, the standard procedure of finding adiabatic Born-Oppenheimer (BO) wavefunctions is not necessarily the most appropriate starting point. Indeed, viewed in this framework, electronic transitions (which can occur in various ranges of internuclear distances R) are induced by non adiabatic dynamical terms involving the radial and rotational nuclear velocities ($- \nabla_R$). These terms might be so large that the original assumptions of the BO approximation breakdown. In fact, since the problem generally amounts to solve a set of coupled equations in a given subspace, any expansion basis in this subspace could equally well be used. The problem then turns out to be that of defining a meaningful reduced set of states accounting for the largest part of the interactions and amenable to pratical calculations. The problem of defining diabatic states is to be placed in this context. In practice, when choosing the basis of electronic wavefunctions one should particularly consider :
    a) the range of internculear distance where the transitions are

R. Daudel, A. Pullman, L. Salem, and A. Veillard (eds.), Quantum Theory of Chemical Reactions, Volume I, 1-23.
Copyright © 1979 by D. Reidel Publishing Company.

susceptible to occur,
     b) the relative importance of the different interactions in this
range (e.g. the various Hund's cases),
     c) the radial velocity of nuclei in the transition region.

     The following examples will provide some illustrations of these
ideas and are intented to sample the general lines along which we are
developing the research in this field.

DIABATIC STATES AND INELASTIC PROCESSES IN HE^+ + HE COLLISIONS

     The He^+ + He collision offers a large variety of excitation mecha-
nisms and the methods developed to handle these processes are illustra-
tive of some aspects of the molecular treatment of atomic collision.

     It was first pointed out by Lichten[3] that the two lowest $^2\Sigma_g^+$ and
$^2\Sigma_u^+$ BO adiabatic states of the He_2^+ molecule ion could not account for
the large resonant charge transfer process of the He^+ + He collision at
moderate energies. The area between the corresponding potential energy
curves, entering the charge transfer probability as

$$P_{CT} = \frac{1}{2} \sin^2(\frac{1}{v} \int (E_{\Sigma g} - E_{\Sigma u}) \, dR)$$

was indeed found to be too small. This situation led Lichten[3] to suggest
another approach based on single configuration states built from molecular orbitals (MO). These states were termed diabatic states. In that
picture (Fig. 1), the $^2\Sigma_u^+$ and the $^2\Sigma_g^+$ states issuing from He^+(1s) +
He($1s^2$) have respectively the $(1s\sigma_g^2 \, 2p\sigma_u)$ and the $(1s\sigma_g \, 2p\sigma_u^2)$ configura-

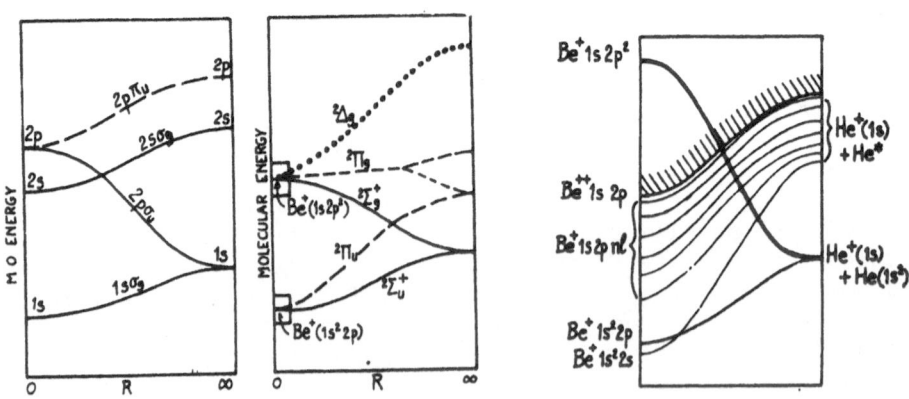

Fig. 1 Scheme of the correlation diagrams for the MO and the
        states of the He^++He system. The MO diagram also illus-
        trates the general correlations of the low-lying orbi-
        tals in a symmetric system.

tions. As R decreases and tends to zero, the $(1s\sigma_g \; 2p\sigma_u^2)^2\Sigma_u^+$ core excited
state correlates with the $1s2p^2$ auto-ionizing configuration of the Be^+
ion. On its way in, the corresponding energy curve crosses the infinite
$(1s\sigma_g^2 \; n l\sigma_g)^2\Sigma_g^+$ Rydberg series as well as the associated continuum
$(1s\sigma_g^2 \; \phi_{k\varepsilon l} \sigma_g)^2\Sigma_g^+$. These crossings (later called diabatic II crossings[4])
occur around the distance where the MO energy differences (ε_i) are such
that (Fig. 2)

$$\varepsilon 2p\sigma_u - \varepsilon 1s\sigma_g \simeq \varepsilon n l\sigma_g - \varepsilon 2p\sigma_u$$

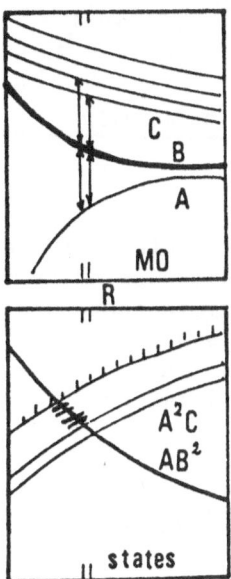

Fig. 2. General scheme for diabatic II crossings[4]. A vacancy
is present in the inner MO (A) and gives rise to the
core excited configuration state (AB^2). This state
crosses the Rydberg series (A^2C) when :
$$2\varepsilon_B \simeq \varepsilon_A + \varepsilon_C$$
This situation always occurs in rare gas$^+$ + rare gas
collisions.

According to the Wigner-Von Neumann[5] theorem these crossings will
appear as avoided crossings in the adiabatic BO representation. Consequently the area between the $^2\Sigma_g^+$ and the $^2\Sigma_u^+$ energy curves, correlated
with the $He(1s) + He(1s^2)$ level, is smaller when calculated with the
adiabatic curves than with the diabatic curves. The failure of the adia-

batic approach to account for the simple resonant charge transfer process is due to the neglect of higher lying adiabatic states. Indeed, at each avoided crossing (Fig. 3),

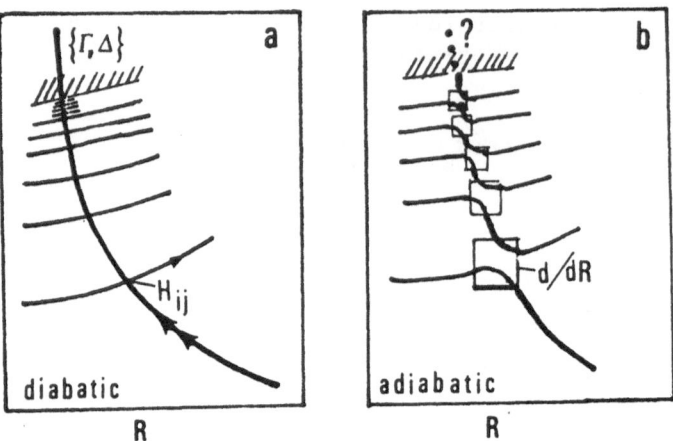

Fig. 3. At each crossing between the diabatic curves (a) the states are weakly coupled by the electronic hamiltonian. At each avoided crossing between the adiabatic curves (b) the states are strongly coupled by d/dR matrix elements. In (a) Γ and Δ are the width and the shift of a level imbedded in a continuum.

the adiabatic wave function displays rapid variations with R, as the configuration changes from e.g. $(1s\sigma_g\ 2p\sigma_u^2)$ to $(1s\sigma_g^2\ nl\sigma_g)$. Huge non adiabatic couplings $(<i|d/dR|j> \simeq 15\ a_0^{-1})$ between the $^2\Sigma_g^+$ states then require inclusion of all higher Rydberg states as well as the associated continuum ! (Fig. 3). The treatment of elastic and resonant charge transfer scattering would be completely intractable in such a representation. On the other hand, the single configuration diabatic states $(1s\sigma_g\ 2p\sigma_u^2)$ $^2\Sigma_g^+$ and $(1s\sigma_g^2\ nl\sigma_g)\ ^2\Sigma_g^+$ cancel all of the d/dR, d^2/dR^2, L^+L^-, matrix elements as shown in ref. 6. The only remaining interaction is due to interelectronic repulsions (electronic correlation) :

$$<(1s\sigma_g\ 2p\sigma_u^2)|\ \sum_{i<j} 1/r_{ij}\ |(1s\sigma_g^2\ nl\sigma_g)> \lesssim 3.10^{-2}\ \text{Hartree}$$

Comparison of the couplings, in the adiabatic $(v_R <i|d/dR|j>)$ and diabatic $(<i|H_{el}|j>)$ representations, shows that for all nuclear radial velocities larger than $v_R \simeq 2.10^{-3}$ a.u. the diabatic single configuration description has to be prefered. This representation actually enables the description of the system as undergoing essentially elastic scattering slightly perturbed by inelastic transitions at curve crossings. This, in turn, allows the use of two state approximations (as e.g. the simple Landau-Zener-Stueckelberg approximation[7], the distorted wave approximation[6] and two-state quantal close coupling treatments[8] for treating excitation, charge transfer and ionization processes of the type :

$$He^+(1s) + He(1s^2) \rightarrow He^+(1s) + He(1sn\ell)$$
$$\rightarrow 2He^+(1s) + e^-$$

at a few hundred eV collision energies. Such electronic transitions at curve crossings are known to yield oscillations in the differential cross sections arising from the interference between the waves scattered by the two accessible potentials inside the curve crossing (Fig. 3).

As the collision energy reaches the keV energy range, experiments[9,10] show the selective single and double excitation of the n=2 levels of He.

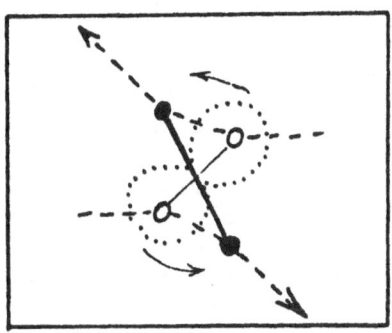

Fig. 4. Schematic view showing that after the sudden rotation of the internuclear axis (O → ●) the σ MO becomes a mixture of σ and π MO's.

Following an idea of Bates and Williams[11], McCarrol and Piacentini[12], invoked a rotatinal coupling at small R between states involving the closely lying $2p\sigma_u$ and $2p\pi_u$ MO's (Fig. 1) namely :

$$(1s\sigma_g\ 2p\sigma_u^2)^2\Sigma_g^+ - (1s\sigma_g\ 2p\sigma_u\ 2p\pi_u)^2\Pi_g - (1s\sigma_g\ 2p\pi_u^2)^2\Delta_g \ ;$$
$$(1s\sigma_g^2\ 2p\sigma_u)^2\Sigma_u^+ - (1s\sigma_g^2\ 2p\pi_u)^2\Pi_u$$

Rotational coupling is known to occur at small impact parameters (b) and small R, where the rapid rotation of the internuclear axis lets no time for the electrons to readjust adiabatically. A σ MO, before the rotation of the internuclear axis, then becomes a mixture of σ and π MO's, after the axis flipping (Fig. 4). This situation yields a σ ↔ π transition which is induced by the component of the electronic angular

momentum perpendicular to the collision plane (Ly). Semi-classically
this coupling writes :

$$v_o b \ <\sigma|Ly|\pi>/R^2,$$

v_o being the relative nuclear collision velocity.

The $2p\pi_u$ MO which correlates with $\pi_u 2p$ atomic orbitals (AO), Fig.
1, thus provides a selective route for populating the n=2 levels of He.

Only recently have the potential energy curves of the He_2^+ molecule
been calculated[13], for small R, to handle this problem. In order to
avoid the linear dependances appearing at small R among LCAO-MO wavefunctions, a one-center expansion[6,14,15] was used and joined to the
two-center results at larger R ($\geq$ 0.5 a$_o$). Actually, such a procedure,
which might introduce discontinuities, can be avoided as recently discussed by Gauyacq[16]. It suffices to add united atom AO's to the two-
center expansion basis. When, for example, the $1s_A$ and $1s_B$ orbitals
become almost linearly dependent the ungerade combination can be deleted
provided the proper 2p AO's for the united atom have been included in
the expansion basis on both the two centers.

To treat the excitation process via rotational coupling at small
R it might seem, at first sight, that a set made of 2-u states and a set
of 4-g states would suffice. Furthermore, since the two ($1s\sigma_g$ $2p\sigma_u$ $2p\pi_u$
$^2\Pi_g$ states can be built in such away that only one of them, namely
$(1s\sigma_g(2p\sigma_u 2p\pi_u)^1\Pi_g)^2\Pi_g$, couples rotationally with the incident
$(1s\sigma_g 2p\sigma_u^2)^2\Sigma_g^+$ state only three states in the gerade set would suffice.
Actually this view is too naïve since it does not take into considera-
tion an important drawback of the single configuration description based
on the LCAO-MO approach in symmetric systems. Since a g or u orbital
dissociates on both of the two He centers, one gets in the dissociation
of e.g. the considered $(1s\sigma_g(2p\sigma_u 2p\pi_u)^1\Pi_g)^2\Pi_g$ state a mixture of the
following separated-atom states :

$$He_A^+ + He_B^{*}(1s\ 2p\ ^3P)$$
$$He_A^+ + He_B^{*}(1s\ 2p\ ^1P)$$
$$He_A^{++} + He_B^-(1s^2\ 2p\ ^2P)$$
$$He^{+*}(n=2) + He(1s^2\ ^1S)$$

and the same with interchange of A and B (see similar examples in ref.[4,
6,15,16]). This means that there exist three other configuration state
functions :

$$(1s\sigma_g(2p\sigma_u 2p\pi_u)^3\Pi_g)^2\Pi_g$$
$$(1s\sigma_g^2\ 3d\pi_g)^2\Pi_g$$
$$(2p\sigma_u^2\ 3d\pi_g)^2\Pi_g$$

which interact at large R with the considered $^2\Pi_g$ state (Fig. 5).

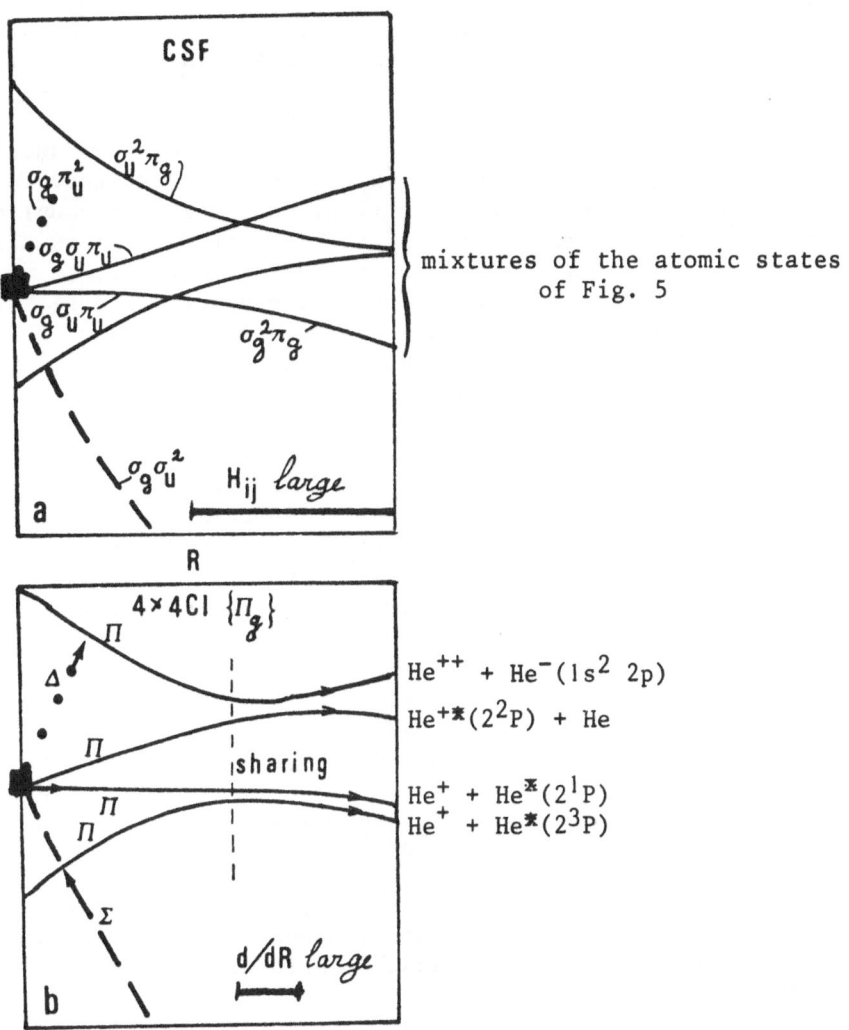

Fig. 5. Schematic dissociation and couplings of some $^2\Pi_g$ states involved in the rotational coupling mechanism of the He$^+$ + He system. (CSF stand for configuration state functions).

The same difficulty occurs with the $^2\Pi_u$ and the $^2\Delta_g$ states. This situation prevents from extracting the scattering S matrix once the primary rotational excitation of the $^2\Pi_g$, $^2\Delta_g$ and $^2\Pi_u$ states has been solved. It might thus seem that either of a 4-fold configuration interaction

(CI) in each of the considered symmetries, or a valence bond (VB) description would be more appropriate at large separations. Unfortunately, the VB approach suffers from similar defects to those discussed above, but at small R. Furthermore, if CI wavefunctions are used, non adiabatic d/dR couplings appear in each of the above 4-fold subspaces reflecting the change of the wavefunctions from a molecular (LCAO) description to an atomic (VB) description (Fig. 5). Consequently, there exists a range of R values where none of single configuration or CI wavefunctions can reduce the problem to less than $4-^2\Pi_g$, $4-^2\Delta_g$, $4-^2\Pi_u$ states. In practice, however, at relatively low velocities, non adiabatic coupling might become negligible and the major part of the excitation will go adiabatically into one of the states in each symmetry. In contrast, at very high velocities the populated states will suddenly project on to all of their atomic components. In the intermediate regime under consideration, the initial population will be more or less shared between the various atomic components of the type discussed above. Therefore, all of the state involved in the dissociation region have to be considered[13].

In order to avoid the sharply peaking d/dR matrix elements arising in the adiabatic CI approach, it suffices to integrate the scattering-closed-coupled equations in the single configuration LCAO-MO representation up to a sufficiently large value of R. In this representation, the states have no d/dR coupling matrix elements since they differ by at least two spin orbitals[13,15,17]. Although, at large R, such states give rise to large off diagonal Hel matrix elements, these couplings tend to constant values in the assymptotic region. Consequently, in this region, constant linear combinations of the diabatic states yield the proper non interacting atomic states without introducing any cumbersome d/dR contribution. In practice, this transformation is carried out during the integration of the scattering equations, just before extracting the S matrix[18]. The Gordon integration method[19,20], actually enables this procedure, to be easily achieved. Indeed, since in this method the equations are locally transformed, at each integration step, from the diabatic to the (local) adiabatic representation and then back to the diabatic one, it suffices to stay in the adiabatic basis once R gets large enough[18].

Fig. 6 shows the results of the above procedure in the treatment of the He(1s 2p ^{3,1}p) excitation via rotational coupling in He$^+$ + He.

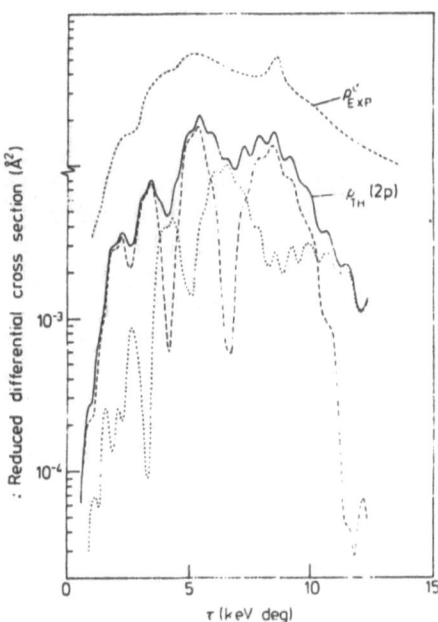

Fig. 6. Reduced differential cross section[18], $\rho = \theta \sin\theta \, \sigma(\theta)$
as functions of the reduced angle $\tau = E\theta$, for the
$2^1P(---)$ and $2^3P(--)$ direct excitations of He in
$He^+ + He$ collisions at $E_{Lab} = 2$ keV. The sum of these
theoretical cross sections $\rho TH(-)$ is to be compared
with experiment ρ_{EXP}.

This procedure has also been applied by Gauyacq[21] to the study of
the single and double excitations of levels involving the 2s and 2p orbitals of He in He + He collisions at keV energies. In addition, the
dissociations into unstable negative ions of the type $He^-(1s^2 \, 2p)$ and
into doubly excited states were considered[21] in some detail in order to
assess the contribution of such species to the ionization processes of
this system.

CHARGE TRANSFER PROCESSES IN H^+ + RARE GAS COLLISIONS[22,23] : H^+ + KR.

The experimental work on H^+ + Kr collisions demonstrated that total
charge transfer cross sections amounting to several $Å^2$ could be obtained
at collision energies as low as a few hundred eV. Among the various charge transfer processes the near resonant one :

$$H^+ + Kr(4p^6\ {}^1S_o) \to H({}^2S_{1/2}) + Kr^+(4p^5\ {}^2P_{3/2}) \ ; \quad \Delta E = 0.4 \text{ eV}$$

$$\to H({}^2S_{1/2}) + Kr^+(4p^5\ {}^2P_{1/2}) \ ; \quad \Delta E = 1.06 \text{ eV}$$

is the most privileged an was recently submitted to detailed experimental differential studies[23].

In order to provide the basis for a molecular study of this process, we have performed a set of calculations on some states of the $(H-Kr)^+$ molecule ion[24]. Since single configuration wavefunctions usually provide a good starting point, SCF calculations were first carried out on the ground state $(\ldots 8\sigma^2)X^1\Sigma^+$ configuration and some singly as well as doubly excited states were determined in the virtual orbital approximation. CI calculations limited to $28-{}^1\Sigma^+$ states were also performed in order to obtain a more reliable ground state potential curve. This potential was subsequently used in low energy $(E < 20$ eV) elastic cross section calculations and quite satisfactory agreement with experiment was achieved[23].

As to inelastic processes, investigation of the coupling term between the $(\ldots 8\sigma^2)\ {}^1\Sigma^+$ ground state (correlated with $H^+ + Kr(4p^6)$) and the lowest $(\ldots 4\pi^3\ 8\sigma^2\ 9\sigma)^1\Pi$ configuration state (issuing from $H(1s) + Kr^+(4p^5))$ showed that rotational coupling should be inefficient in inducing the charge transfer process. Although spin-orbit interactions should be considered to investigate the population of the ${}^2P_{1/2}$ and the ${}^2P_{3/2}$ sublevels of Kr^+, their effect is found to be much too small to have a significant influence on the primary excitation mechanism from the $X^1\Sigma^+$ state. Next, as the $X^1\Sigma^+$ state and the first excited $(\ldots 8\sigma\ 9\sigma)I^1\Sigma^+$ state differ from each other by one MO, a d/dR coupling does exist between the two states :

$$\langle X^1\Sigma^+ | d/dR | I^1\Sigma^+ \rangle = \sqrt{2} \ \langle 8\sigma | d/dR | 9\sigma \rangle$$

(the 8σ and 9σ MO's being respectively correlated with $\sigma 4p_{Kr}$ and $\sigma 1s_H$). Unfortunately the related dynamical coupling $(v_R \langle X | d/dR | I \rangle^r$ amounts at most to 15% of the X-I energy difference at $E = 100$ eV. This is insufficient to fully account for the observed large total charge transfer cross sections at this energy. In addition, since the SCF calculation was performed on the ground state closed-shell configuration, the Brillouin theorem states that no direct coupling is induced by the electronic hamiltonian between the considered X and I states. However, as the X and I states lie very close to each other, at large R, second order interactions should also be considered. Particularly it is found[24] that both the X and the I states predominantly interact with the doubly excited $(\ldots 9\sigma^2)D^1\Sigma^+$ state which dissociates into $H^-(1s^2) + Kr^{++}(4p^4$ $(\sqrt{2}\ {}^1D + {}^1S)/\sqrt{3})$ (Fig. 7). From perturbation theory, it it estimated that the interaction between the X and I states, via the doubly excited $D^1\Sigma^+$ relay state, may become of the same order of magnitude as the X-I energy difference around $R = 6a_o$. Such an interaction at large R can indeed yield an important total cross section and should induce the major part of the charge transfer transitions.

Considering this effect together with the spin-orbit interaction (see further discussions below) in a four-state quantal close coupling treatment, we obtained qualitative agreement with the measured differential cross sections[23] and only about 60% of the total experimental charge transfer cross section[25]. Hence, our neglect of the radial dynamical coupling was not completely justified since it resulted in significant discrepancies in the comparison with experiment.

However, the calculation of the d/dR matrix element as a function of R is an expensive task requiring twice as much effort as that necessary to determine the energy curves. In addition, there exist, to our knowledge, no efficient programs to handle such a coupling (involving first derivative terms) in quantal close coupling calculations. Furthermore, the introduction, in our approach, of the $D^1\Sigma^+$ relay state, which is negligibly populated, means that we have failed in adequately discribing the process. Intuitively, a more appropriate description would confine the interactions to the lowest $^1\Sigma^+$ states without calling for any extra (virtual contribution).

As the charge transfer transitions occur at large R (6-7a$_o$), it seems that a description in terms of atomic orbitals in a valence bond (VB) approach could provide a better starting point[26]. Considering the VB states issuing from $H^+ + Kr$ and $H + Kr^+$, one has respectively the following configuration state functions :

$$|..... \sigma 4p_{Kr} \ \overline{\sigma} 4p_{Kr}|$$

$$1/\sqrt{2}\{|..... \sigma 4p_{Kr} \ \overline{\sigma} 1s_H|-|... \ \overline{\sigma} 4p_{Kr} \ \sigma 1s_H|\}$$

In this two-fold subspace one may choose two arbitrary orthonormal states. Using Schmidt orthogonalization we can build the two functions :

$$X^1\Sigma^+ = |...\sigma 4p_{Kr} \ \overline{\sigma} 4p_{Kr}|$$

$$\tilde{Y}^1\Sigma^+ = \mathcal{N} [1/\sqrt{2}\{|...\sigma 4p_{Kr} \ \overline{\sigma} 1s_H|-|...\overline{\sigma} 4p_{Kr} \ \sigma 1s_H|\}- \sqrt{2} \ \Sigma_i <\phi^i_{Kr}|\sigma_{1s_H}>\tilde{X}^1\Sigma^+]$$

$$= 1/\sqrt{2}\{|...\sigma 4p_{Kr} \ \overline{\sigma} 1\tilde{s}_H|-|...\overline{\sigma} 4p_{Kr} \ \sigma 1\tilde{s}_H|\}$$

where : $\sigma 1\tilde{s}_H = \mathcal{N} (\sigma 1s_H - \Sigma <\phi^i_{Kr}| \ \sigma 1s_H> \ \phi^i_{Kr})$, $\mathcal{N}$ being a normalizing factor and ϕ^i_{Kr} being a Kr groundstate AO.

The $\tilde{Y}^1\Sigma^+$ state thus involves a projected AO of the type introduced by O'Malley[27] in his definition of diabatic states. Therefore the states can be called projected valence bond (PVB) states[24]. Considering the d/dR matrix element between the above two states one has

$$<\tilde{Y}|d/dR|\tilde{X}> = \sqrt{2} \ <\sigma 1\tilde{s}_H|d/dR|\sigma 4p_{Kr}>$$

$$=-\sqrt{2} \ <\sigma 4p_{Kr}|d/dR|\sigma 1\tilde{s}_H>$$

As the d/dR operation is carried out in keeping fixed the electron
coordinates refered to the center of mass of the nuclei (CMN) and
since, in addition, this origin can be considered, to a good approximation, to lie on the Kr nucleus, one gets :

$$\langle \hat{I} | d/dR | \hat{X} \rangle \simeq 0$$

One then also meets with Smith's definition of diabatic states[28]. The
remaining coupling between the $\hat{X}$ and the $\hat{I}$ states is then only due to
the matrix element $\langle \hat{X} | H_{el} | \hat{I} \rangle$ which decreases exponentially at large R.
The energy difference between the $\langle \hat{X} | H_{el} | \hat{X} \rangle$ and the $\langle \hat{I} | H_{el} | \hat{I} \rangle$ potential
energy curves (Fig. 7) remains almost constant down to $R \simeq 4a_o$ and it
becomes equal to twice the $\langle \hat{X} | H_{el} | \hat{I} \rangle$ coupling at $R \simeq 7a_o$. The latter
distance defines the charge transfer transition region[29]. These fea-
tures are those usually assumed in the charge transfer models that appeared in the literature (see e.g. ref. 26). Proceeding further and defining the $^3\Pi$ and the $^1\Pi$ states issuing from $H(1s) + Kr^+(4p^5)$ in the same
way as was done for $\hat{I}^1\Sigma^+$, it is readily seen that none of these Π states
couples directly with the incident $\hat{X}^1\Sigma^+$ state.
In addition, the indirect interaction, via the doubly excited state
dissociating into $H^-(1s^2) + Kr^{++}(4p^4)$, is found to be two orders of
magnitude smaller than that considered in the LCAO-SCF approach. Consequently charge transfer primarily takes place between the two $^1\Sigma^+$-PVB
diabatic states : $\hat{X}$ and $\hat{I}$.

To check the accuracy of the PVB approach, one can perform a CI
calculation involving a set of singly and doubly excited PVB states
and compare the results with the CI calculations involving the LCAO-SCF
basis set. Very close agreement is found down to $R \simeq 4a_o$. Below this
distance, although the energy separation between the two lowest PVB +
CI $^1\Sigma^+$ states is correctly reproduced, the shape of the curves is incorrect (Fig. 7). This effect is readily understood since the frozen orbitals involved in the PVB configuration state functions can hardly account
for molecular distorsion and polarization[24]. These effects can, of course, be accounted for by a larger CI than was actually performed. Although
such a defect should have minor consequences in the calculation of total
charge transfer cross sections, it is expected to dramatically affect
the interference patterns in differential cross sections, especially
rainbow effects. However, as the PVB approach provides an accurate, adequate and convenient description at large R ($\gtrsim 4a_o$), it can be used
there to treat the charge transfer process. At smaller R, since the
coupling between the $\hat{X}$ and $\hat{I}$ states gets much larger than their energy
difference and since the PVB treatment is deficient, a more appropriate
description would be based on non interacting adiabatic states (of e.g.
the LCAO-SCF + CI type). The latter approach, as small R, enables to
properly account for interference patterns[23]. The transformation from
one description to the other is carried out at a distance R_D (Fig. 7)
during the integration of the scattering equations and is very similar
to the procedure described for the $He^+ + He$ system[18].

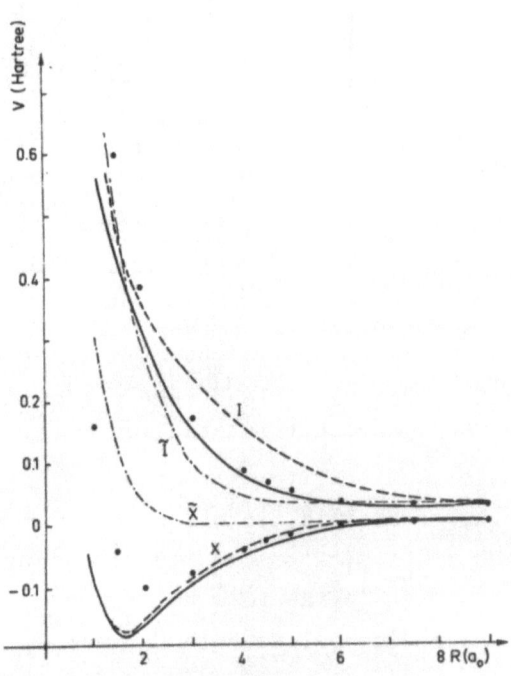

Fig. 7. Scheme of the two lowest $^1\Sigma^+$ potential energy curves
of the (H-Kr)$^+$ molecule ion[24] : --- SCF curves
— LCAO-SCF+CI curves
-•- PVB single configuration curves
• PVB + limited CI
R_D defines the distance where the basis change from
the adiabatic states to the diabatic states is performed
in the course of the cross section calculations[23].

As for spin-orbit effects, which are (of course) included in our
cross section calculations, they are also handled within a diabatic
representation. We actually considered $\Omega = 0^+$ states[24]. In a similar
manner as that discussed for the dissociation of the states in the He$_2^+$

system, we can choose one of the three following approaches (Fig. 8) :

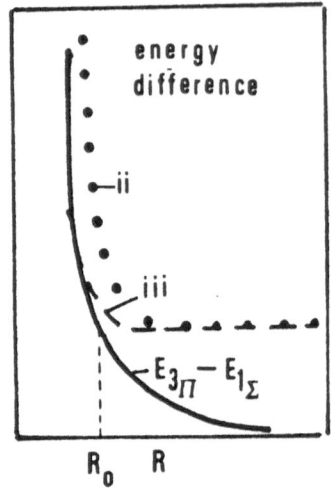 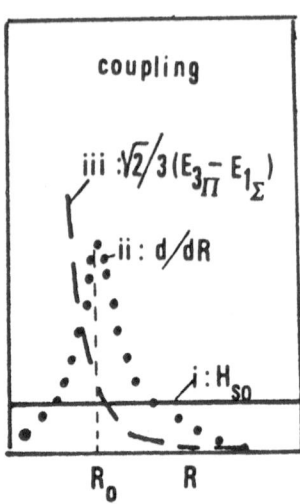

Fig. 8. Energy differences and coupling schemes in the three
        possible representations that may be used for descri-
        bing the spin-orbit interaction[30] (see text).

(i)   a molecular approach, at small R, in which the spin effects (H_{so})
      are dominated by the electrostatic effects. In this approach the
      appropriate states are the $|^1\Sigma^+>$ and the $|^3\Pi>$ states correlating
      with $H+Kr^+(4p^5)$

(ii) adiabatic states which diagonalize both H_{el} and H_{so} and introduce
      a d/dR coupling.

(iii) an atomic approach, at large R, where spin effects dominate. The
      appropriate states are then constant linear combinations of the
      previous $|^1\Sigma^+>$ and $|^3\Pi>$ states which diagonalize H_{so}.

The adiabatic representation (ii) continuously goes from case (i) to
case (iii). In both cases (i) and (iii) the states are diabatic since they
involve no d/dR coupling (Smith's definition[28]). Again as in the case of
the $He^+ + He$ collision a basis change from case (i) to case (iii) can be
performed in the assymptotic region. However, since in the present system the transition occurs at large R, such a basis change is needless and
only the representation of case (iii) is sufficient. Of course, for
physical significance the basis change can be performed when actually
needed.

     Both the differential (Fig. 9) and total charge transfer cross
sections calculated using the above method agree satisfactorily with
experiment. The same procedure has been applied to the $H^+ + Xe$ collision
[31] and was again found to sucessfully account for the experimental findings[22]. This method appears to be very promising and could be extended

with great ease to other systems even if they involve lighter targets than those considered above[23,32].

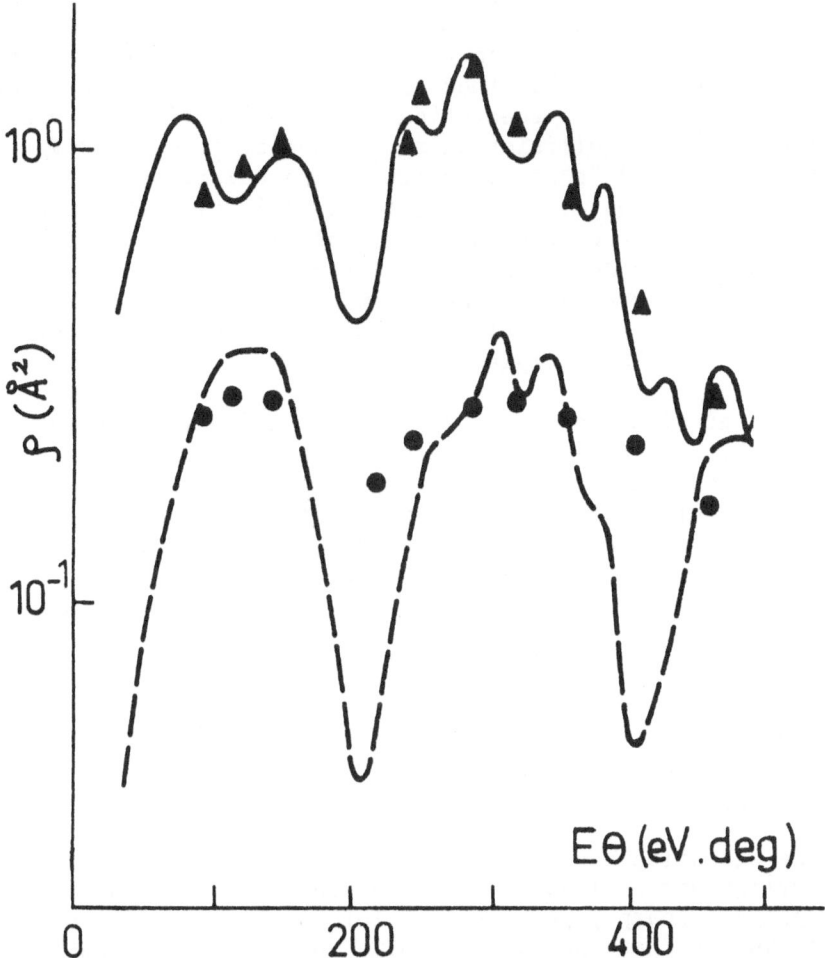

Fig. 9. Reduced differential cross sections for the near resonant charge transfer processes of the H^+ + Kr system[23] at E_{Lab} = 100 ev.
H^+ + Kr → H + Kr^+($^2P_{3/2}$) : ⎯ theory, ▲ experiment
→ H + Kr^+($^2P_{1/2}$) : -- theory, ● experiment.

INELASTIC PROCESSES IN Li^+ + HE COLLISIONS AND DIABATIC QUASI U-G CROSSINGS.

The Li^+ + He collision provides the simplest case of an asymmetric closed-shell + closed-shell interaction and has been the subject of a number of experimental works[33,34,35]. The theoretical analysis of elastic

and inelastic processes of this system has then motivated a few studies
on the ground and the excited molecular states of the (Li-He)$^+$ molecule.
Le Sech et al.[36,37] invoked the similarity of this system with He-He
and consequently studied the excitation mechanisms induced by rotational
coupling using perturbed united atom (B$^+$) wave functions. Later, Junker
and Browne[38] reported impact parameter scattering calculations involving
adiabatic CI wavefunctions, but they neglected two-electron excitation
processes. In order to study theoretically the Li$_K$ excitation processes
observed recently[39], at collision energies ranging from few keV to several keV, we carried out[40] a set of ab initio calculations on several
singly and doubly excited molecular states of (Li-He)$^+$. The considered
states were those involving the MO's that correlate with the 1s, 2s and
2p AO's of either Li or He. We incidentally found some anomalies in the
excited curves reported by Junker and Browne[38] which prompted us to
reinvestigate the one- and two-electron excitation mechanisms of the
Li$^+$ + He collision at keV energies.

One of the difficulties that arises in studying this system is the
occurence (R $\lesssim$ 0.5 a$_o$) of a set of pseudo crossings in its Σ[38] and Π
potential energy curves, even in the single configuration description
of the states[40]. This feature is easily understable by analyzing the
behavior of the MO's. The MO correlation diagram of the (Li-He)$^+$ molecule
should be similar to that of the neighboring He-He isoelectronic system
(Fig. 1). However, since the (Li-He)$^+$ system has no more an inversion
symmetry, the 2sσ_g-2pσ_u crossing is actually avoided (Fig. 10a). This
avoided MO crossing (called diabatic I^4) reflects itself in all the potential energy curves that involve the second and the third SCF-MO's :

$$(1\sigma^2\ 2\sigma^2),\ (1\sigma^2\ 2\sigma\ 3\sigma),\ (1\sigma^2\ 3\sigma^2)\ {}^1\Sigma^+$$
$$(1\sigma^2\ 2\sigma\ 1\pi),\ (1\sigma^2\ 3\sigma\ 1\pi)\qquad {}^1\Pi$$

The same situation occurs for the 2σ and 3σ MO's of the NeO system
studied by Briggs and Taulbjerg[41]. In that case the <2σ|d/dR|3σ> matrix
element displays a sharply peaking behavior and maximizes at a value
of $\sim$200 a$_o^{-1}$. Such a sharp peaking is quite cumbersome and should be
avoided by looking for equivalent diabatic MO's that smooth and reduce
this coupling. This could be achieved by an R dependent rotation of the
2σ and 3σ MO's as proposed in ref. 41. However, this procedure is rather
expensive since it consists first in calculating the d/dR matrix elements
which, in turn, requires a relatively large set of SCF calculations at
short intervals in R. A more attractive method, based on the same idea,
has recently been proposed by Gauyacq[42], Instead of working on the mini-
mization of the d/dR coupling, he requires that a large number of H$_{el}$
and/or L$\pm$ matrix elements, which vary rapidly around the avoided crossing, be smoothed by the rotation. An alternative procedure, involving
the shielded diatomic (SDO) method[43] has been used by Aubert and Le Sech
[44] to handle a similar situation in the (He H)$^+$ molecule ion. We were
interested instead in a more straightforward method that could yield
directly the diabatic MO's by taking advantage of the quasi-symmetric

character of the system considered. Our SCF calculations showed that
the 2σ-3σ avoided crossing occurs at R ≃ 0.275 a$_o$ (Fig. 10a). At such
small internuclear distances, close to the united atom limit, it is
natural to think and it is actually verified (Fig. 10b), that the 2σ
and the 3σ MO's have already acquired, outside of the pseudo crossing,
a quasi-u or a quasi-g character.

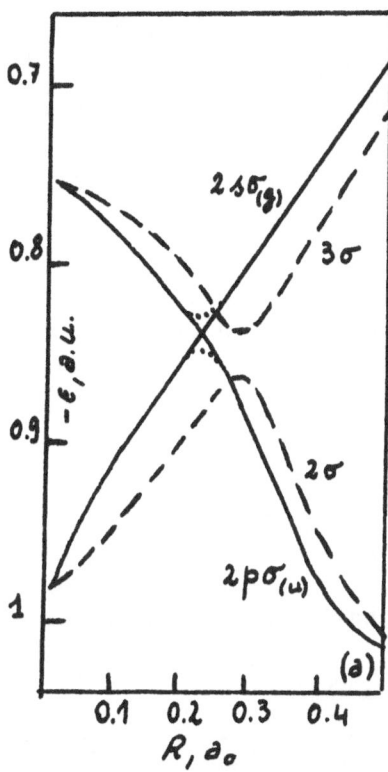

Fig. 10a. Scheme of the 2nd and 3rd σ MO's of the (Li-He)$^+$
system in its lowest $^3\Sigma^+$ state[40]
-- standard adiabatic SCF MO's
— diabatic SCF-MO's related to the ficitious
symmetric zero order hamiltonian H^g_o.

The different location of the crossing distance
is due to the use of different LCAO expansion
basis sets (this feature can of course be improved)

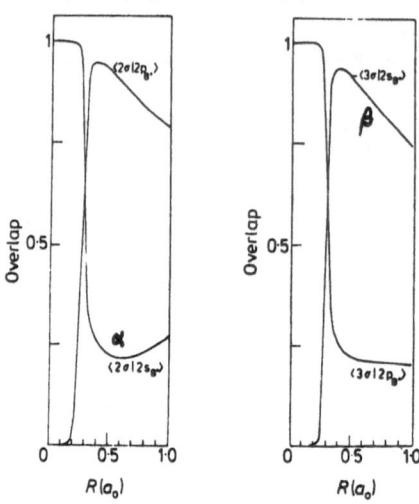

Fig. 10b. Variation in the character of the 2σ and 3σ adiabatic
SCF MO's (-- in fig. 10a) as determined from the
overlap with the B^+ united atom orbitals

$$\alpha = <2\sigma|2s_B{}^+>$$
$$\beta = <3\sigma|2s_B{}^+>$$

Consequently, the zeroth order hamiltonian, that will serve to generate
the diabatic MO's, should take this underlying symmetry into account.
Writing the electronic hamiltonian as a sum of gerade and ungerade terms:

$$H_{el} = H_o^g + V^u + Z_{He} Z_{Li}/R$$
$$\text{where}: H_o^g = \sum_i -\frac{1}{2}\Delta j - \frac{2.5}{r_{Lij}} - \frac{2.5}{r_{Hej}} + \sum_{i<j}\frac{1}{r_{ij}}$$
$$V^u = \sum_j -\frac{0.5}{r_{Lij}} + \frac{0.5}{r_{Hej}}$$

we are insured that the H_o^g operator will generate non interacting $2s\sigma_g$
and $2p\sigma_u$ MO's that cross (Fig. 10a). On the other hand, the small V^u g
term (which vanishes as $R \to 0$) will induce $2p\sigma_{(u)}$ - $2s\sigma_{(g)}$ transitions.

Due to the different masses of Li and He a small d/dR coupling
remains :

$$<2s\sigma_{(g)}|d/dR|CMN|2p\sigma_{(u)}> = \frac{1}{2}(\frac{M_{Li} - M_{He}}{M_{Li} + M_{He}}) <2s\sigma_{(g)}|\frac{\partial}{\partial z}|2p\sigma_{(u)}>$$

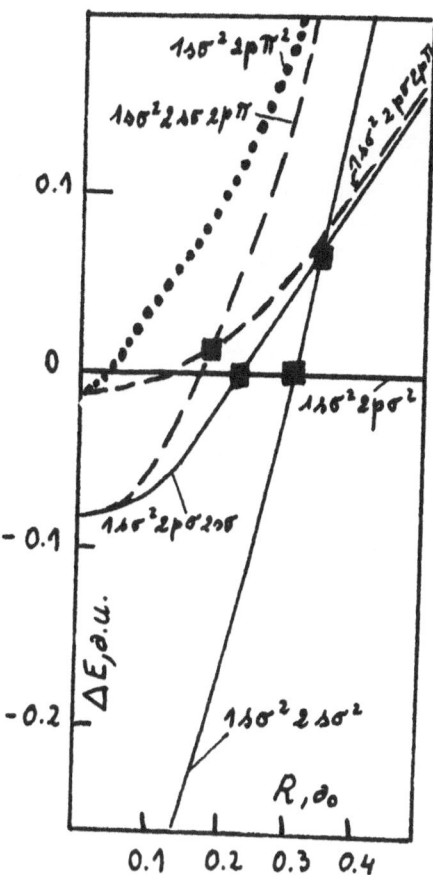

Fig. 11. Diabatic single configuration states of the (Li-He)$^+$ system[40]. All the crossings shown (■) are generated by the $2s\sigma$-$2p\sigma$ diabatic I[4] MO crossing of Fig. 10a.

In the above formula the mass factor amounts to 0.227 and z is measured from the midpoint of the internuclear axis. The above term can now be evaluated analytically using prolate spheroïdal coordinates $(\xi = (r_{Li} + r_{He})/R, \eta = (r_{Li} - r_{He})/R)$:

$$\frac{\partial}{\partial z} = \frac{2}{R(\xi^2 - \eta^2)} \left(\eta(\xi^2 - 1) \frac{\partial}{\partial \xi} - \xi(1 - \eta^2) \frac{\partial}{\partial \eta} \right)$$

We are currently involved in the process of evaluating how large this remaining interaction is.

Preliminary quantal cross section calculations at E_{Lab} = 2 keV

neglecting this coupling and involving the 6 configuration states of
Fig. 11 show satisfactory agreement with experiment (Fig. 12) and im-
prove among previous theoretical investigations.

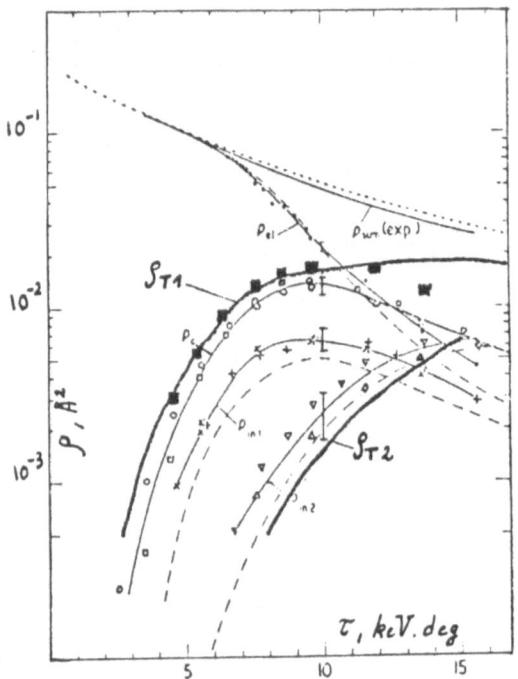

Fig. 12. Theoretocal one and two electron-excitation (—) cross
sections of the Li$^+$ + He collision at E_{Lab} = 2 keV. ρ_{33}
is the sum of the experimental data $\rho_c + \rho_{inl}$
—— experimental data of ref. 33
-- experimental data of ref. 34
... purely elastic classical cross section[33]

The same procedure has also been used for the Na$^+$-Ne system[45] to
generate the quasi u,g diabatic I $4f\sigma$–$3s\sigma$ MO crossing[4,46,47]. In this
case, however, the outer shells get their quasi-symmetric characters at
larger R values than the more compact inner shells. This problem has
been circumvented by using two sets of MO's. The inner MO's are obtained
from a standard SCF calculation whereas the valence and Rydberg orbitals
are obtained in the same way as described for the Li$^+$-He system. The
latter orbitals are next orthogonalized to those in the first set. This
step does not destroy the u or g properties of the outer MO's (Fig. 13).

It is worth mentioning that an important outstanding problem has

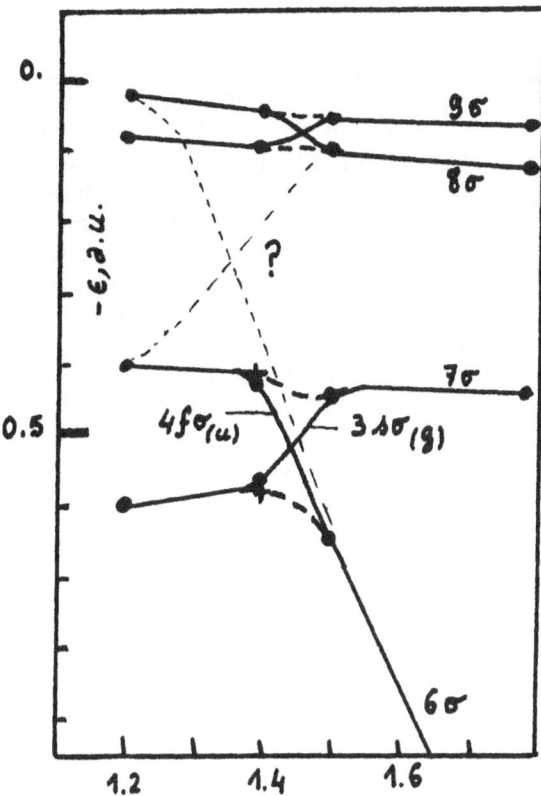

Fig. 13. Outer-shell diabatic (—) and adiabatic (--) SCF MO's for the Na-Ne$^+$ system in its $^3\Sigma^+$ state. The two higher lying MO's are virtual unoccupied orbitals. The question mark indicates the region where the as yet unsolved $\sigma_{(u)} - \sigma_{(u)}$ crossing should lie.

to be overcome for the primary excitation mechanisms, at relatively small R, to be completely solved. The problem is that of treating diabatic I crossings[4] between MO[46,47] that have no underlying symmetries as those discussed above. The $4f\sigma_u - 4p\sigma_u$, $4f\sigma_u - 5p\sigma_u$ crossings of the Ne-Ne system[46] are typical examples of this open problem. An interesting approach to this problem is now being developed by Gauyacq[42]. This approach is based on the slowly varying character required for diabatic MO's and involves frozen orbitals determined outside of the pseudo crossing region. The attractive SDO method[43] of Aubert and Le Sech[44] which will have to prove itself in heavy systems could provide an alternative procedure to handle this problem.

CONCLUSION

The few cases discussed here illustrate some typical processes encountered in atom-atom collision problems and some of the theoretical methods developed to deal with them. Of course, the list is by no means complete and it was only intended to sample some trends in this field. It has particularly been emphasized that one can get ridof the "sacro-sanct" B.O. adiabatic representation and still achieve an adequate and convenient description of the electronic part of an atom-atom collision problem. Or particular significance are the two seemingly contradictory aspects that have been developed here, namely, the molecular continuously adjusting states and the atomic or almost frozen description. These features stem from the inherent specific characteristics of atom-atom collisions at moderate velocities. Indeed, at such velocities the motion of the nuclei is sufficiently slow for a molecular aspect to build in and it is sufficiently rapid for some "characters" (atomic characters, node conservation rule[47], conservation of the configuration type, conservation of some underlying symmetric, etc....) to be preserved.

REFERENCES

1 Centre de Mécanique Ondulatoire Appliquée, 23, rue du Maroc, 75019 Paris, France.
2 Laboratoire des Collisions Atomiques, Université Paris XI, bât 351, 91405 Orsay, France.
3 Lichten, W.: 1963, Phys. Rev. 131, p. 229.
4 Brenot, J.C., Dhuicq, D., Gauyacq, J.P., Pommier, J., Sidis, V., Barat, M. and Pollack, E.: 1975, Phys. Rev. A, 11, p. 1245.
5 Von Neumann, J. and Wigner, E.P.: 1929, Physik Z., 30, p. 467.
6 Sidis, V.: 1973, J. Phys. B: Atom. Molec. Phys. 6, p. 1188.
7 Olson, R.E.: 1971, Phys. Rev. A 4, p. 1030.
8 Evans, S.A. and Lane, N.F.: 1973, Phys. Rev. A 8, p. 1385.
9 Barat, M., Dhuicq, D., François, R., McCarroll, R., Piacentini, R.D. and Salin, A.: 1972, J. Phys. B : Atom. Molec. Phys. 5, p. 1343.
10 Barat, M., Brenot, J.C. and Pommier, J. : 1973, J. Phys. B : Atom Molec. Phys. 6, p. L 105.
11 Bates, D.R. and Williams, D.A.: 1964, Proc. Phys. Soc. (London) 83, p. 425.
12 McCarroll, R. and Piacentini, R.D.: 1971, J. Phys. B : Atom. Molec. Phys. 4, p. 1026.
13 Stern, B., Gauyacq, J.P. and Sidis, V. : 1978, J. Phys. B 11, p. 653.
14 Briggs, J.S. and Hayns, W.R.: 1973, J. Phys. B : Atom. Molec. Phys. 6, p. 514.
15 Sidis, V., Dhuicq, D. and Barat, M.: 1975, J. Phys. B : Atom. Molec. Phys. 8, p. 474.
16 Gauyacq, J.P.: 1976, J. Phys. B : Atom. Molec. Phys. 9, p. 2289.
17 Sidis, V. and Lefebvre-Brion, H.: 1971, J. Phys. B : Atom. Molec. Phys. 4, p. 1040.
18 Stern, B., Gauyacq, J.P. and Sidis, V..: 1978, J. Phys. B 11, p. 653.

19 Gordon, R.G.: 1969, J. Chem. Phys. 51, p. 14.
20 Gordon, R.G.: 1971, Meth. Comput. Phys. 10, p. 81.
21 Gauyacq, J.P.: 1976, J. Phys. B : Atom. Molec. Phys. 9, p. 3067.
22 Kubach, C., Benoit. C., Sidis, V.. Pommier, J. and Barat, M: 1976 J. Phys. B : Atom. Molec. Phys. 9, p. 2073.
23 Benoit, C., Kubach, C., Sidis, V., Pommier, J. and Barat, M.: 1977 J. Phys. B : Atom. Molec. Phys. 10, p. 1661.
24 Kubach, C. and Sidis, V.: 1976, Phys. Rev. A. 14, p. 152.
25 Koopman, D.W.: 1967, Phys. Rev. 154, p. 79.
26 Demkov, Yu. N: 1964, Sov. Phys. J.E.T.P. 51, p. 322.
27 O'Malley, T.F.: 1969, J. Chem. Phys. 51, p. 322.
28 Smith, F.T.: 1969, Phys. Rev. 179, p. 111.
29 Stueckelberg, E.C.G.: 1932, Helv. Phys. Acta 5, p. 369.
30 Sidis, V., Invited Papers and Progress Reports of the ninth International Conference on the Physics of Elect. and Atom. Collisions (Ed. J.S. Risley and R. Geballe, Univ. of Washington press)(1975).
31 Kubach, C. and Sidis, V.: 1976, J. Phys. B 9, p. L413.
32 Kubach, C., Thèse de Doctorat d'Etat, 1976 (Univ. Paris-Sud)
33 François, R., Dhuicq, D. and Barat, M.: 1971, J. Phys. B : Atom. Molec. Phys. 5, p. 963.
34 Lorents, D.C. and Conklin, G.M.: 1971, J. Phys. B : Atom. Molec. Phys. 5, p. 950.
35 Park, J.T., Pol, V., Lawler, J., George, J., Aldag, J., Parker, J. and Peacher, J.L.: 1975, Phys. Rev. A, 11, p. 857.
36 Le Sech, C., McCarroll, R. and Baudron, J.: 1973, J. Phys. B : Atom. Molec. Phys. 6, p. L 11.
37 Le Sech, C., Thèse de Doctorat d'Etat, Université Paris XI (1976).
38 Junker, B.R. and Browne, J.C.: 1974, Phys. Rev. A 10, p. 2078.
39 Stolterfoht, N. and Leithauser, U.: 1976, Phys. Rev. Letters 36, p. 186.
40 Sidis, V., Stolterfoht, N. and Barat, M.: 1976, 2nd Int. Conf. on Inner Shell Ionization Phenomena (Freiburg), p. 68. 1977, J. Phys. B 10, p. 2815.
41 Taulbjerg, K. and Briggs, J.S.: 1975, J. Phys. B. : Atom Molec. Phys. 8, p. 1895.
42 Gauyacq, J.P., 1978, J. Phys. B 11, p. L217.
43 Aubert, M., Bessis, N. and Bessis, G.: 1974, Phys. Rev. A 10, p. 51 and 1974, Phys. Rev. A 10, 61.
44 Aubert, M and Le Sech, C: 1976, Phys. Rev. A 11, p.
45 Sidis, V., Barat, M., Pommier, J. and Agusti, J., work in progress.
46 Fano, U. and Lichten, W.: 1965, Phys. Rev. Letters 14, p. 627.
47 Barat, M. and Lichten, W.: 1972, Phys. Rev. A 6, p. 211.

RECENT THEORETICAL DEVELOPMENTS IN THE DYNAMICAL STUDY OF MECHANISTIC DETAILS IN ORGANIC REACTIONS

Xavier CHAPUISAT and Yves JEAN

Laboratoire de Chimie Théorique (ERA 549)
Université de Paris-Sud, Bâtiment 490,
Centre d'Orsay, 91405 ORSAY Cedex

FOREWORD

Dynamical studies in Chemistry (either experimental or theoretical) were the matter of many publications in the past ten years (see ref. [1] for a review). More recently the scope of Chemical Dynamics was extended to organic reactions. The eldest theoretical studies in this field were limited to simplified models of reactions of small molecules (such as are $CH_3NC \rightarrow CH_3CN$ [2], $K+RI \rightarrow KI+R$ where $R=CH_3$ or C_2H_5 [3] , $T+CH_4 \rightarrow TH+CH_3$ or $H+CH_3T$ [4], etc...). The potentials used were empirical. The first organic reaction studied with a semi-empirical (CNDO2) potential was $CH_2+H_2 \rightarrow CH_4$ by Wang and Karplus [5]. Finally the optical and geometrical isomerizations of cyclopropane were investigated from a dynamical point of view by Jean and Chapuisat, using an ab initio potential [6]. All these studies were undertaken within the framework of classical mechanics which is known to describe correctly the atomic motions [7].

In the latter study mentioned above, a suitable formulation of classical mechanics for large molecules in which many degrees of freedom are constant (as are the models developed by quantum chemists to compute the potentials) is introduced [1]. It will be referred to as the Theory of Constrained Systems. The main results are as follows (q denotes a vector whose components are the actual generalized coordinates of the potential and $\dot{q}$ and p are respectively the velocities and conjugate momenta). The constraints find expression in the fact that the kinetic energy is a non diagonal quadratic form of either $\dot{q}$ or p, and the coefficients depend on q. In matrix notation [1] it is a dot product, either $T = 1/2<\dot{q}|A(q)|\dot{q}>$ or $T = 1/2<p|A^{-1}(q)|p>$ where A(q), the socalled constraint matrix, is mass dimensioned :

$$A_{jk}(q) = \sum_{i=1}^{N} m_i \left[\frac{\partial x_i}{\partial q_j} \frac{\partial x_i}{\partial q_k} + \frac{\partial y_i}{\partial q_j} \frac{\partial y_i}{\partial q_k} + \frac{\partial z_i}{\partial q_j} \frac{\partial z_i}{\partial q_k} \right] \qquad (0\text{-}1)$$

Here, the atoms are N ; the ith one, whose cartesian coordinates x_i, y_i and z_i depend on q, weights m_i. In matrix notation the equations of motion are :   (i) $\dot{q} = A^{-1}(q)p$ (0-2)   and

$$\dot{p} = -\nabla\{1/2<p|A^{-1}(q)|p> + V(q)\} \qquad (0\text{-}3)$$

25

R. Daudel, A. Pullman, L. Salem, and A. Veillard (eds.), Quantum Theory of Chemical Reactions, Volume I, 25–52.

in the Hamilton formulation (∇ denotes the gradient operator ($\partial/\partial q$))
and (ii) $\ddot{\underset{\sim}{q}} = -A^{-1}(q) \underset{\sim}{S}$ (0-4) in the Lagrange formulation. Here

$$\underset{\sim}{S}(q,\dot{q}) = \nabla\{1/2<\underset{\sim}{q}|A(q)|\underset{\sim}{q}> - V(q)\} + \underset{\sim}{R} \qquad (0-5)$$

and $\underset{\sim}{R}$ is a vector the jth component of which is :

$$R_j = \underset{k,1}{\Sigma} \frac{\partial A_{jk}(q)}{\partial q_1} \dot{q}_k \dot{q}_1 \qquad (0-6)$$

Applications of the Theory of Constrained Systems are presented in
references [1] and [6].

In the present contribution, we intend to present a new application
(part 1) and an improvement (part 2) of the Theory of Constrained
Systems. In part 1, a dynamical model study of optical and geometrical
isomerizations of 1,1,2,2-tetramethylcyclopropane, including all the
main degrees of freedom (ring opening and rotations of the terminal
groups) is presented. A wide range of initial conditions is studied and
a qualitative comparison of the dynamical behaviour of this molecule
with that of unsubstituted cyclopropane is proposed. Three conclusions
emerge : (i) heavy substituents considerably slow down the internal
rotations (s) within the diradical ; (ii) all the reactive trajectories
involve only a single rotation of 180° of one or both terminal groups
before ring-closure ; (iii) for the substituted molecule, the positions
of the transition states are more important factors than their heights.
The connection of this study with experiment is discussed. In part 2,
coming back to the basic formalism of the Theory of Constrained Systems,
we propose to treat within an adiabatic framework some coordinates of
secondary importance from a dynamical point of view but that are nevertheless used as optimization parameters in computing the potentials.

1. SUBSTITUENT EFFECTS IN ISOMERIZATIONS OF 3-MEMBERED RINGS

1.1. Introduction

The geometrical and optical isomerizations of cyclopropane [8]
most probably involve a diradical intermediate, the 1,3-propane-diyl[9].
The masses of substituent groups attached to reactive centers seem to
influence in a crucial way the ratio of reaction rates for the different
possible reactive processes within the diradicals. Thus the ratio
k_{cyc}/k_{rot} of the cyclisation rate constant (k_{cyc}) to the terminal group
rotation rate constant (k_{rot}) increases rapidly with increasing substituent masses. The ratio is 0.26 for 1-methyl, 2-ethylcyclopropane [10]
and 11 for tetramethylcyclopropane-d_6 [11]. Substituting heavy groups
to hydrogen atoms disfavours the process which involves rotations. This
effect agrees with the statistical thermodynamic argument stated by
Benson [12] : the larger the moment(s) of inertia of the rotor(s), the
more negative the entropy of activation.

In a previous article [6b] we presented the results of a dynamical
study of isomerization reactions of unsubstituted cyclopropane. Total

energy thresholds were obtained for the different isomerization processes as well as reactive and non-reactive bands of initial vibra-
tional-torsional energy in the terminal groups. The aim of the present
section is (i) to study the way in which the previous results are
modified when the masses of the terminal groups are increased substan-
tially ; (ii) to connect our dynamical results with the experimental
results. The masses are chosen in order to simulate the case of 1,1,2,2-
tetramethylcyclopropane. All other parameters are unchanged, i.e. we
conserve the same dynamical model (cf. ref. 6b , fig. 10) and the same
potential energy surface (cf. ref. 6b , eq. 6) as previously. The
consequences of the latter crude working hypothesis will be discussed
along with the results.

The discussion presented below on the nature of the reaction product depends on the basic assumption that only the substituted bond
breaks [10,11,13]. However, in order not to condition our thinking, it
should be kept in mind that this assumption was never demonstrated to
be definitively valid. In particular, in isomerizations of deuterated
molecules, the three carbon-carbon bonds break equally. Here it is
known from experiments that concerted rotations of two methylene groups
are always the dominant mechanism [8]. Indeed when the doubly substituted

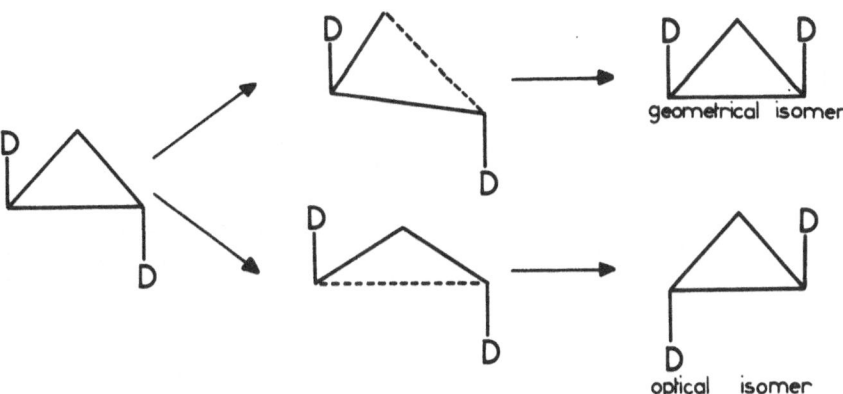

Figure 1 - For deuterated cyclopropane, both optical and
geometrical isomers are formed through concerted rotations.
The nature of the reaction product just depends on the
doubly substituted bond being broken, or not.

bond breaks, the optical isomer obtains whereas the geometrical isomer is produced if one of the two singly substituted bonds breaks (see Fig. 1).

1.2. Equation of motion and initial conditions

The equations of motion are formally the same as that of the previous study (cf. ref. 6b , eqs. 11-19). The only modifications concern the mass m_1 which is equal to 15 (methylene group) instead of 1 (hydrogen atom), the parameter

$$\rho = \frac{M + 2m_1}{M + 2m_2} = 3$$

and the length 1 which changes from 1.08 Å (CH-bond length) to 1.59 Å (distance of carbon atom to the center of gravity of an adjacent methyl group).

All the calculated trajectories start from the molecule in its equilibrium conformation (ring angle : $2\alpha° = 60°$; plane of the terminal bonds perpendicular to the plane of the carbon ring : $\theta_1° = \theta_2° = 0$).The initial conditions include the total energy (E_{tot}), the fraction of E_{tot} in the vibrations of the terminal $C(CH_3)_2$ groups ($E_{rot}°$) and the way in which $E_{rot}°$ is distributed among the two groups treated as rigid rotors (angle $\delta°$ such that $tg\delta° = \dot{\theta}_1°/\dot{\theta}_2°$, where the $\dot{\theta}$'s are rotational velocities (cf. ref. 6b , eqs. 20-22)).

It should be noticed that 1,1,2,2-tetramethylcyclopropane is actually a very complicated reacting molecule (57 internal degrees of freedom). The effects of the low-lying bending modes, for example, may play an important role in influencing the dynamics. This is not taken into account in the present model.

1.3. Results

Total energy thresholds for the different types of isomerization. In table I we give the computed total energy thresholds for respectively unsubstituted cyclopropane [6b] and tetrasubstituted cyclopropane. Specific values are given for the three basic motions : concerted conrotatory ($\delta° = + 45°$), rotation of a single group ($\delta° = 0$) and concerted disrotatory ($\delta° = - 45°$) respectively. These results show that :

i) The excess total energy which is necessary for completion of reactive trajectories is much greater for tetrasubstituted cyclopropane than for unsubstituted cyclopropane. At first sight, this result is all the more surprising. As a matter of fact it is known that the isomerization of substituted cyclopropane requires a smaller activation energy than that of unsubstituted cyclopropane [11, 14]. In fact, our working hypothesis of the same potential energy surface being used for the dynamics of both cyclopropane and tetramethylcyclopropane, makes the

$\delta°$	T. S.[a] energy (position)	thresholds for tetrasubstituted cyclopropane	thresholds[a] for unsubstituted cyclopropane
+ 45°	59.8 (58°)	65.2	60.4
0°	61.6 (90°)	65.9	62.3
- 45°	61.9 (50°)	64.6	62.1

(a) : ref. [6b]

Table I – Potential energies and positions of the transition states, and total energy thresholds, for both cyclopropane [6b] and tetramethylcyclopropane. The three basic motions are respectively concerted conrotatory ($\delta°=+45°$), rotation of a single group ($\delta°=0$) and concerted disrotatory ($\delta°=-45°$). The energies reported here are in kcal/mole, within a scale where the origin is the potential energy of cyclopropane in its equilibrium conformation.

direct comparison between experimentally measured activation energies and the present theoretical energy thresholds invalid. Indeed the decrease of the activation energy observed experimentally when substituting the molecule is due to a strong destabilization of the reactant molecule because of steric repulsion. Our static potential energy surface does not take into account this strong destabilization. Thus our dynamical results cannot reproduce the experimentally observed ordering of the activation energies of the overall reactions. However they show clearly that the various rotational processes involved in the reaction mechanism are always more difficult when cyclopropane is substituted. Indeed trajectories with $E_{tot} < 64.6$ kcal/mole are all unreactive. They are characterized by an opening of the ring immediately followed by reclosure prior to any possible passage over a transition state (see Figure 2). This is due to the very slow rotations of heavily substituted terminal groups. For instance the rotational velocity of a $C(CH_3)_2$-group is, for a given rotational energy, six times slower than that of a CH_2-group [6a] 2.

ii) Compared with the unsubstituted case the order of appearance of the three basic reaction processes – as E_{tot} increases – is reversed. In the present case the first reaction process is a synchronous disrotatory

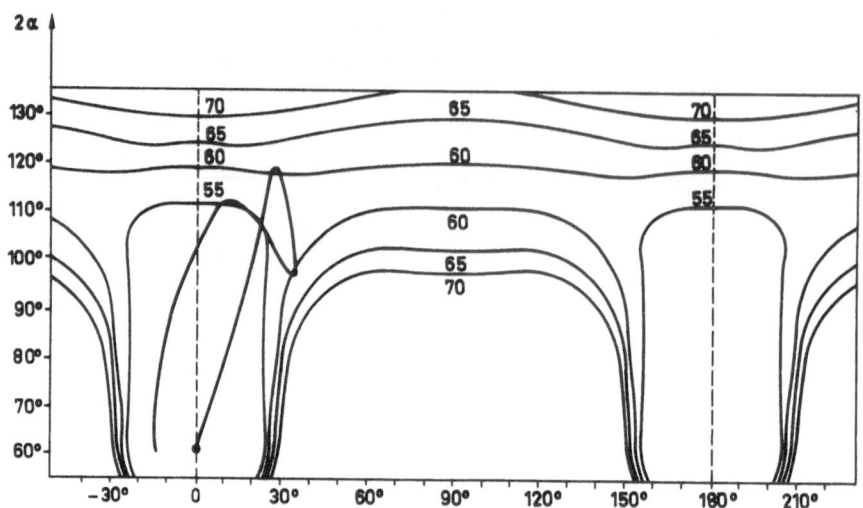

$E_{TOT} = 62$ k cal/mole _ $E^o_{ROT} = 14$ k cal/mole _ $\delta° = 45°$

Figure 2 - A typical non reactive conrotatory trajectory.
Although the total energy is largely in excess of the
potential energy of the transition state, the transition
state cannot be attained because it lies at too great a
rotational angle for both terminal groups. The rotational
velocity of these groups is not sufficient.

motion (threshold at 64.6 kcal/mole). Now, rather strikingly, this
motion corresponds to the highest possible transition state (T.S. at
61.9 kcal/mole). In fact, for competing transition states which differ
by only 1 or 2 kcal/mole, it seems that the position of the cols is
crucial in the case of heavily substituted molecules. Since the rotations
are very slow the smaller the rotation leading to a given transition
state conformation, the more likely the passage of a trajectory over this
col [3] (cf. Table I). This statement is illustrated in Figure 3. Identical initial conditions lead to a non reactive trajectory when $\delta°=+45°$
(conrotatory motion, cf. Figure 3a) and to a reactive trajectory when
$\delta°=-45°$ (disrotatory motion, cf. Figure 3b). The transition state for
disrotatory motion is more easily attainable than that for conrotatory
motion although the former is a bit higher in energy [15] [4] because it
corresponds to a smaller rotational angle (50° instead of 58°). Moreover,
the entrance valley for ring opening is wider, and its corner into the
upper rotational valley less sharp, for the disrotatory motion. This
allows the representative point of a disrotatory trajectory to "cut the
corner" more easily, because of centrifugal forces. This particularity
is of secondary importance for cyclopropane. It becomes important for
tetramethylcyclopropane.

Reaction times and dynamical behaviour. The isomerization process

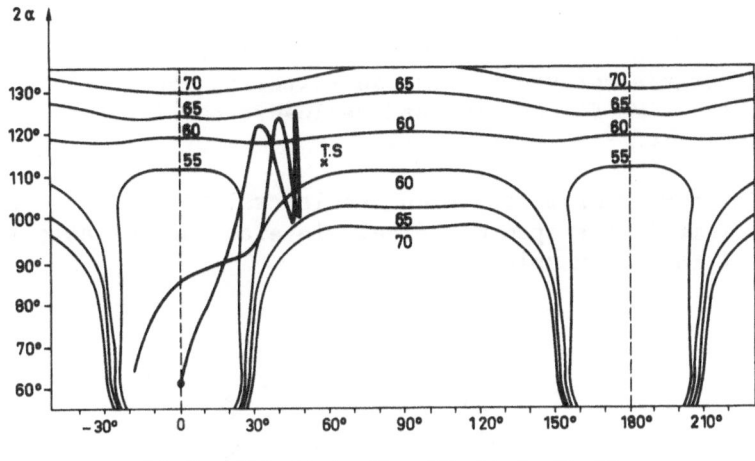

(a) $E_{TOT} = 65$ k cal/mole $-$ $E_{ROT}^{0} = 18$ k cal/mole $-$ $\delta^{0} = 45°$

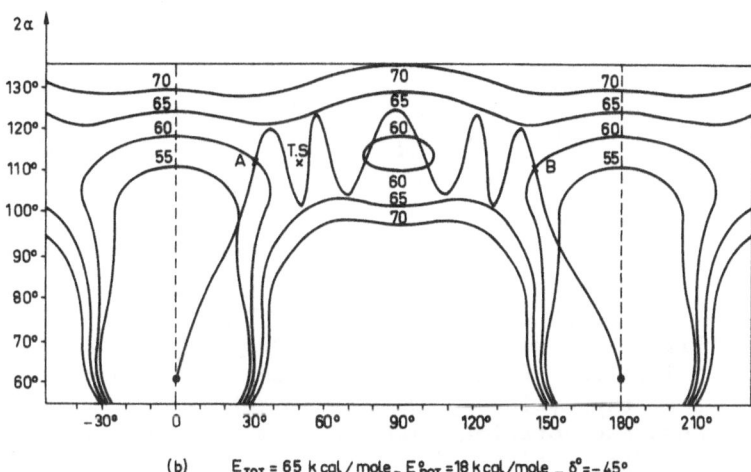

(b) $E_{TOT} = 65$ k cal/mole $-$ $E_{ROT}^{0} = 18$ k cal/mole $-$ $\delta^{0} = -45°$

Figure 3 – Passage over a given transition state (T.S.)
depends more on its position than on its height. Thus
identical initial conditions lead to (a) a non-reactive
trajectory if $\delta^{0}=+45°$ (conrotatory T.S. : 59.8 kcal/mole
at $\theta_1=\theta_2=58°$) and (b) a reactive trajectory if $\delta^{0}=-45°$
(disrotatory T.S. : 61.9 kcal/mole but at $\theta_1=-\theta_2=50°$).

as a whole lasts noticeably longer for a substituted cyclopropane molecule than for the non-substituted one (about four times longer). Thus
in Figure 3b the time interval for isomer formation is 9.1×10^{-13}
second, more than three quarters (7.5×10^{-13} second) of which are used
for the rotation of the terminal groups from $\theta=30°$ (point A in Figure
3b) to $\theta=150°$ (point B). Thus the opening and reclosure processes last
about 1.6×10^{-13} second. On the average, a reactive process lasts about

10^{-12} second for tetramethyl-cyclopropane compared with 2.5×10^{-13} second for cyclopropane [6b]. The latter is typically decomposed in 1.2×10^{-13} second for the opening and the reclosure and 1.3×10^{-13} second for the rotations. Thus substitution does not modify significantly the time for opening and reclosure but lengthens by a factor of 6 the rotations.

The number of rebounds on the lateral walls of the upper rotational valley (within the open diradical) is also considerably increased (see Figure 4 for an example).

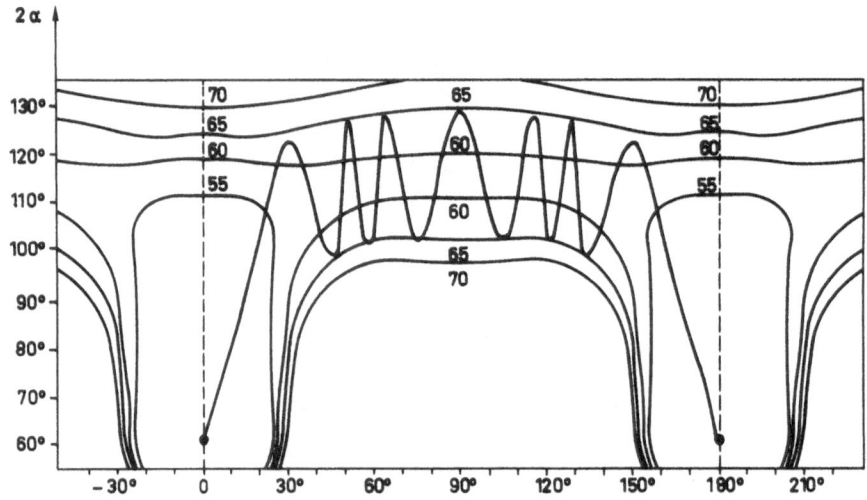

$E_{TOT} = 66 \text{ k cal / mole} \_ E^\circ_{ROT} = 14 \text{ k cal / mole} \_ \delta^\circ = 45^\circ$

Figure 4 - Rebounds on the lateral walls of the upper rotational valley of the diradical species increase in number in tetramethylcyclopropane because substituent methyl groups slow down the rotational motion more than the ring-opening (or closure) motion.

It should be noticed that the existence of heavy substituents on the reactive carbon atoms results in the disappearence of all the trajectories that were characterized by several full 180° rotations of the terminal groups in the previous study. Such trajectories are observed for cyclopropane, in particular when the excess total energy is large. In the case of tetramethylcyclopropane, the ring recloses very rapidly as soon as the diradical has conformation close to that of the face-to-face diradical, so that no supplementary rotation has time to take place.

Reactive bands. When E°_{rot} is varied for given values of E_{tot} and δ°, there appear alternant reactive and non reactive domains (see Figure 5). In our previous study [6b] we have shown that these "bands" are characterized by the number N of half-vibrations of the terminal groups which take place as the ring-opening motion goes on. If each band

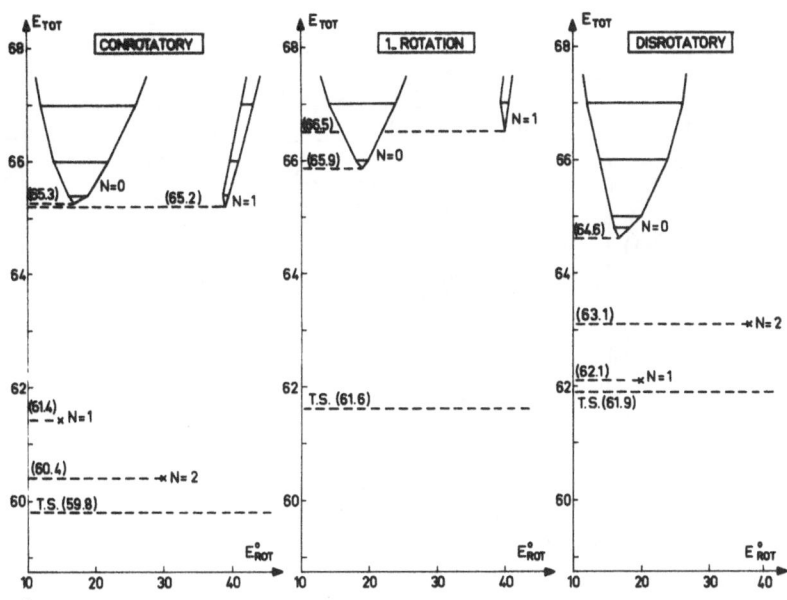

Figure 5 - Reactive bands of tetramethylcyclopropane in the
three basic cases : (a) conrotatory ; (b) 1-rotation and
(c) dirotatory. N is the number of rebounds of the trajectories
on the lateral walls in the entrance ring-opening valley. The
crosses N=1 and N=2 denote for comparison the points where
appear the reactive bands of cyclopropane.

is labeled by the appropriate value of N, the N=1 and N=2 bands are observed in the isomerizations of cyclopropane. In the case of tetramethylcyclopropane the observed bands correspond to N=0 (see Figure 3b) or N=1
(see Figure 6) [5] . The appearance of the N=0 band which is associated
with a direct passage from the ring-opening valley to the upper rotational valley, i.e. with no preliminary bouncing off on the walls of the
entrance valley, is due to the large increase of the mass associated
with the rotational coordinate [16]. The same factor explains why the
N=1-type band is deplaced towards higher initial rotational energies
E_{rot}° (from 18 kcal/mole to 40 kcal/mole on average).

1.4 Discussion

    The results of the present study are qualitatively in good agree-
ment with the experimental results reported at the beginning of this
article. Indeed, for a same set of initial conditions, and once the
ring is opened, the geometrical and optical rotations of the terminal
groups are easier processes for the unsubstituted cyclopropane molecule
than for substituted cyclopropane-type molecules. Since the result we
refer to is the ratio k_{cyc}/k_{rot} it should be noticed that substitution

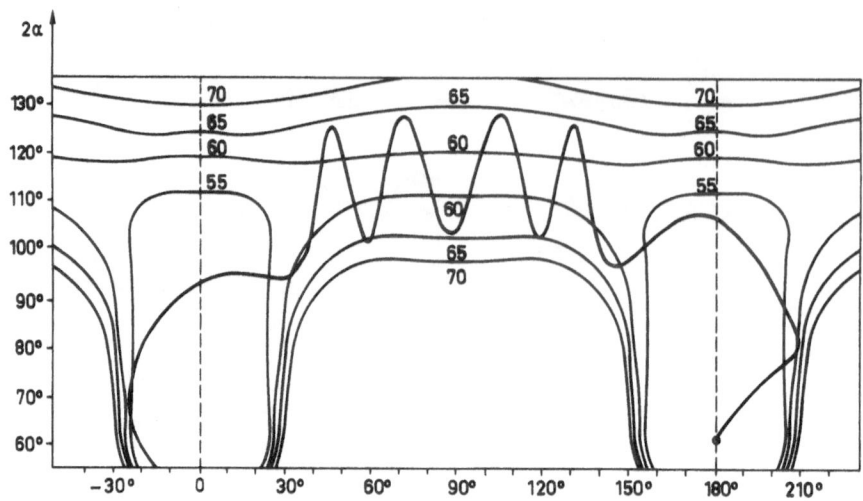

$$E_{TOT} = 66 \text{ k cal/mole} - E^{\circ}_{ROT} = 40 \text{ k cal/mole} - \delta^{\circ} = 45^{\circ}$$

Figure 6 - An example of N = 1 - type trajectory in the conrotatory case.

does not modify significantly the durations of the opening and the reclosure (cf. § 2 above and reference [6a]). Thus substitution of hydrogen atoms on reactive centers by methyl groups disfavours all processes which involve the rotation of the terminal groups and consequantly favours those processes which involve the formation of a bond (the ratio k_{cyc}/k_{rot} increases). Quantitatively the replacement of hydrogens by methyl groups has the same effect as increasing the rotational barriers by about 2.5 kcal/mole in cyclopropane itself.

It should be emphasized, however, that our results cannot be compared with, or related to the experimental measurements of the relative reaction rates for, respectively, the concerted and non-concerted motions in tetrasubstituted cyclopropanes [10,11,13]. This is due to the potential energy surface used throughout the present study which does not reproduce the very strong steric repulsion that exists within the coplanar (edge-to-edge) diradical species formed in these molecules [11,17] (See Figure 7). This strong interaction should have two important consequences : (i) to increase by a large amount the activation energy for concerted rotations, and (ii) to shift the positions of the transition states along the conrotatory and disrotatory pathways on the potential energy surface towards θ=90°. Since, as we have just seen, the actual position of the cols is crucial in determining the probability for dynamical crossover, the quantitative study of the dynamics of heavily substituted cyclopropanes would require to use their true potential energy surfaces.

The major conclusion is that our study provides an example in which the transition state which is the most easily attainable is not that whose potential energy is the lowest. This model indicates that the

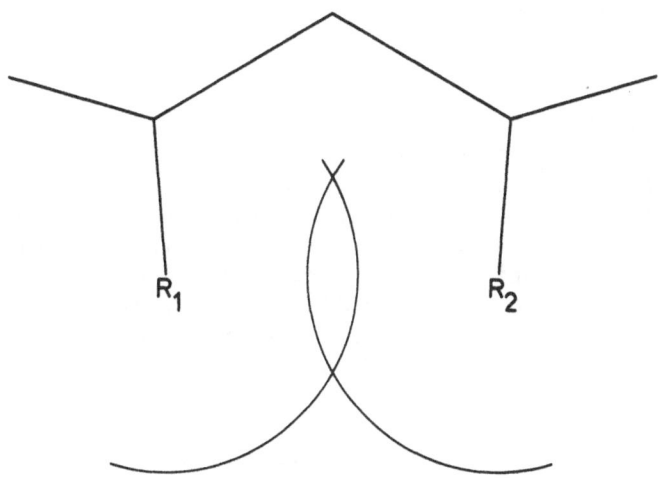

Figure 7 - For heavily substituted cyclopropane the edge-to-edge conformations - in which two substituents are inner groups - are sterically hindered.

rotational phase of the reaction strongly depends on the masses of substituents. It allows us to assert that, if a molecule can isomerize by different processes whose activation energies are close to each other, purely dynamical factors may play an important role in determining the outcome of the reaction [15], in addition to statistical effects. This assertion is by no way a challenge to Transition State Theory [18] since our trajectories are neither sufficiently numerous nor averaged to reproduce the behaviour of the real system.

2. ADIABATIC TREATMENT OF SECONDARY COORDINATES IN CLASSICAL TRAJECTORIES APPLIED TO CONSTRAINED SYSTEMS

2.1. Introduction

The purpose of this section is twofold. First a method is proposed to treat adiabatically certain degrees of freedom of secondary importance in Chemical Dynamics, and second an application of it is presented in a ticklish case. This is an improvement of the Theory of Constrained Systems recently developed by us [1] for the trajectory studies of organic reactions, i.e. for large molecules in which many degrees of freedom are held constant.

This improvement implies an increase of the complexity of the theory at the formal level. It is nevertheless useful as it allows to use in the dynamical trajectories exactly the same model as in the study of the potential, including the optimization variables.

2.2. Adiabatic description of the motion of secondary coordinates.
     Instantaneous adjustement of this to the motion of the main
     coordinates

Most often in Organic Chemistry the reactions involve large
molecules, i.e. numerous atoms. From the dynamical point of view, the
coordinates (positions and momenta) are about six times as numerous.
Thus the complete dynamical treatment is impossible. Indeed obtaining
the equations of motion and solving them stably is difficult. But
above all complete potentials of such systems are definitely unknown.
Most potentials depend on a few coordinates only, even for large
molecules. Thus many degrees of freedom are held constant. In other
words the molecular potentials are known as if the molecules were
constrained systems [1].

Chemical Dynamics in Organic Chemistry aims at studying the
dynamical behaviour of molecules with those degrees of freedom that
are kept in potentials. These will be referred to as independent (or
dynamical) coordinates and denoted by q. However there are often
additional secondary(or adiabatic) coordinates introduced through an
optimization of the potential with respect to them. They are not
variables but rather adjustable parameters. We denote them by Q. They
do not appear explicitly in the adiabaticity corrected potential energy
function used $\mathcal{V}(q)$. Indeed :

$$\mathcal{V}(q) = V(q,Q(q)) \tag{2-1}$$

where $V(q,Q)$ is the unknown complete potential energy function depending
on both q and Q, and $Q(q)$ is a given function (adiabaticity correction).
Thus the situation differs from the case where Q is fixed. Indeed $\mathcal{V}(q)$
and the forces $-\partial \mathcal{V}/\partial q_i$ acting on the dynamical coordinates depend adiabatically on Q. Since dynamics comes along after potential calculations,
we must confide the selection of adiabatic coordinates added to the
independent coordinates by quantum chemists. Thus we look for a formulation of Classical Mechanics for their motion.

In ref. [1] the general expression of the kinetic energy with respect to generalized coordinates (r_i) and velocities $(\dot{r}_i)$ is given (cf.
ref. [1], eqs. 37, 38) :

$$2T = \sum_{i,j=1}^{m} A_{ij}(r_1, r_2 \ldots r_m)\dot{r}_i\dot{r}_j \tag{2-2}$$

where

$$A_{ij}(r_1,r_2\ldots r_m) = \sum_{k=1}^{3N} \frac{\partial R_k}{\partial r_i} \frac{\partial R_k}{\partial r_j} \tag{2-3}$$

is a mass-like coefficient ; it is a matrix-element of the so-called
Constraint matrix $A(r)$. The R's are the natural coordinates as functions of the generalized coordinates. The number of (holonomic) constraints is (3N-3-m). Sharing the generalized coordinates (r) into

independent (q) and adiabatic (Q) coordinates results in a new expression of the kinetic energy and in a new <u>reduced adiabaticity corrected</u> constraint matrix $\overline{A}$ (q) more intricate than A (r).

With the following notations :

a) $\{r_1, r_2 \ldots r_m\} = \{q_1, q_2 \ldots q_n ; Q_1, Q_2 \ldots Q_\ell\}$

r : generalized   q : dynamical   Q : adiabatic
   coordinates   coordinates   coordinates

where :   m = n + ℓ

b)

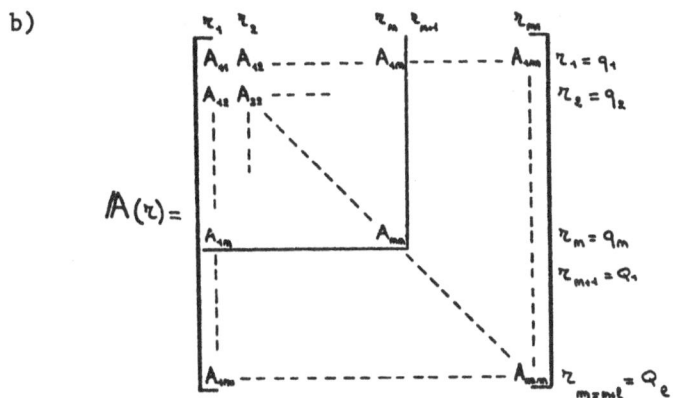

c) Adiabatic coordinates : $Q_i = Q_i(q_1, q_2 \ldots q_n)$ (i = 1,2...ℓ)

d) $\mathcal{A}$ (q) = A (q, Q(q))
the current $\overline{A}$ (q) - matrix element is (i, j = 1, 2 .... ℓ) :

$$(2-4) \qquad \overline{A}_{ij}(q) = \mathcal{A}_{ij}(q) + \sum_{I=1}^{\ell} \frac{\partial Q_I(q)}{\partial q_i} \left[2\mathcal{A}_{j,n+I}(q) + \sum_{J=1}^{\ell} \mathcal{A}_{n+I,n+J}(q) \frac{\partial Q_J(q)}{\partial q_j} \right]$$

Then the adiabaticity corrected kinetic energy is a quadratic function of the only dynamical velocities :

$$2\,\overline{T} = \sum_{i,j=1}^{n} \overline{A}_{ij}(q)\, \dot{q}_j\, \dot{q}_i \qquad\qquad (2-5)$$

Various terms correct $\mathcal{A}_{ij}(q)$ in the expression of $\overline{A}_{ij}(q)$ (see, r.h.s. of eq.2-4). Thus clearly, <u>taking into account adiabatic coordinates modifies the dynamical coupling between the independent coordinates.</u> Hence the motion is altered. In particular the adiabatic coordinates act as energy reservoirs. They can either accept or furnish energy and so modify the motion.

Once the matrix $\overline{A}$(q) is known, the dynamical trajectories obtain in exactly the same way as described in ref. [1]. An example is presented in the next section.

2.3. Dynamical study of isomerizations of cyclopropane. Adiabatic treatment of the pyramidalization motion of the rotating CH_2

In several articles devoted to the transition state for optical isomerization of cyclopropane [19], Salem, Jean et al. obtained a wealth of information on the entire potential energy surface for both geometrical and optical isomerizations (See Figure 8). In particular they distinguished the coordinates of primary importance from that of secondary

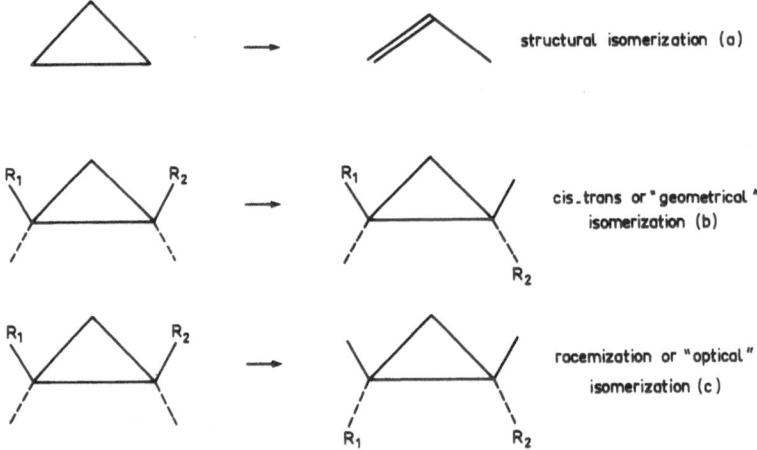

Figure 8 - Definition of the various isomers of a cyclopropane-type molecule.

importance and that to be held constant. The primary coordinates are (i) the ring opening angle (2α) and (ii and iii) the rotation angles (θ_1 and θ_2) of the terminal methylene groups around the adjacent carbon-carbon bonds (See Figure 9).

However an important correction must be introduced. The least-energy states of pyramidalization of the terminal groups change along with reclosure. Indeed the value of β varies from $+30°$ to $-30°$ when the ring recloses in Figure 10 [6a]. Moreover, whenever a terminal group rotates, its pyramidalization must invert algebraically between $\theta=0°$ and $\theta=180°$ through a non-pyramidalized (i.e. trigonal) state at $\theta=90°$ (See Figure 11) [6b]. To obtain the potential energy surface [19], the pyramidalizations β_1 and β_2 of the two terminal CH_2 were adiabatically adjusted to α, θ_1 and θ_2 :

$$\beta_i = \beta_i (\alpha, \theta_i) = \beta_o(\alpha) (1 - 2 \theta_i/\pi) \quad (i=1,2) \qquad (2-6)$$

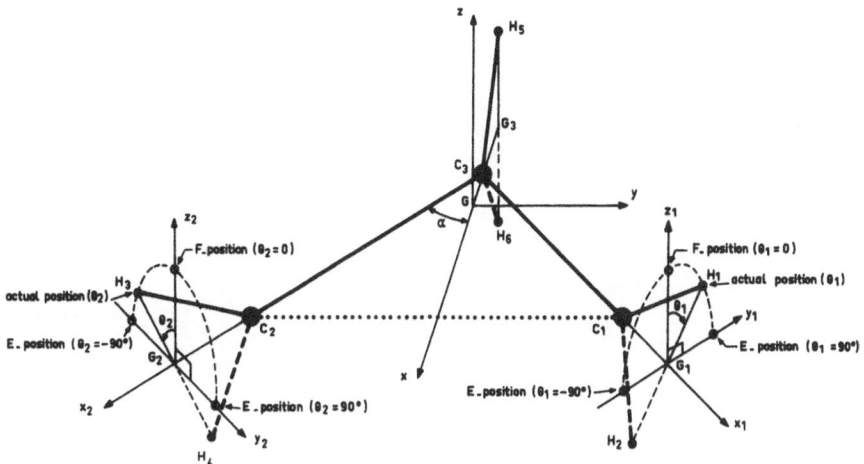

Figure 9 - Definition of the **dynamical** variables α, θ_1 and θ_2 and definition of the **simplified** trigonal (ST) dynamical model.

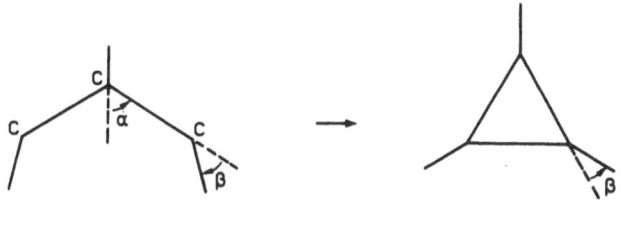

FF ($\beta = +30°$)   CYCLOPROPANE ($\beta = -30°$)

Figure 10 - Definition of the two geometrical coordinates (angles α and β) primarily concerned in the ring closure from the face-to-face (FF) diradical to cyclopropane.

Hence a 3-dimensional adiabaticity corrected potential function obtained:

$$\mathcal{V}(\alpha, \ \theta_1, \ \theta_2) = V\ (\alpha, \ \beta_1(\alpha,\theta_1), \ \theta_1, \ \beta_2(\alpha,\theta_2), \ \theta_2) \qquad (2\text{-}7)$$

The adiabaticity device and the nature of the static surface are described in detail in ref. [6b].

However, although the potential used was adiabaticity corrected, in ref. [6b] the terminal groups remained trigonal throughout the reaction. Thus the kinetic energy was not adiabaticity corrected. It was diagonal in $\dot{\alpha}^2$, $\dot{\theta}_1^2$ and $\dot{\theta}_2^2$ since the planes of rotation of the terminal groups were always perpendicular to the adjacent carbon-carbon bonds, the instantaneous rotation axes. This simplified model was supposed not to result in a severe restriction for dynamics. We test this assumption in the results section below. In order to modify the equations of motion used in ref. [6b] and to include the adiabaticity corrections we first

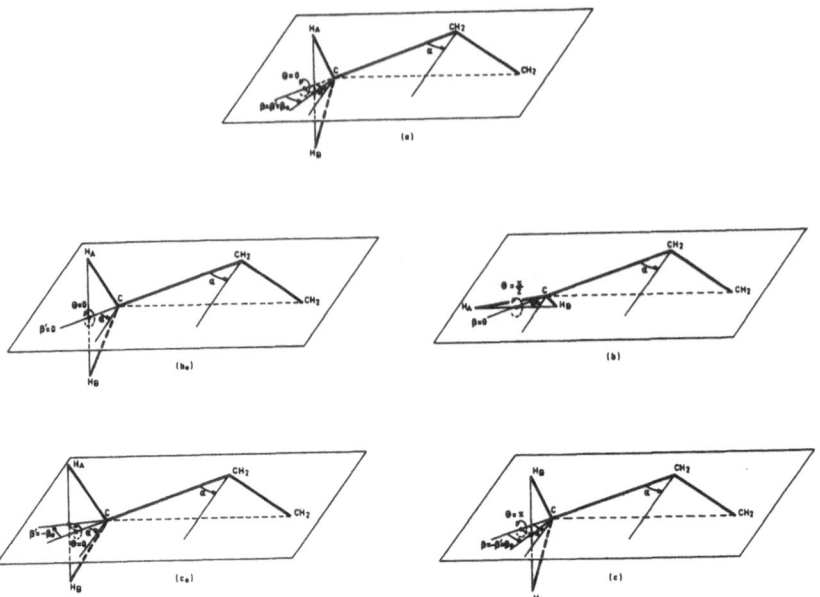

Figure 11 - Pyramidalization inversion of a terminal methylene
group, when rotating around the adjacent carbon-carbon bond.
The true geometries used in the PAC dynamical model and in
computing the potential energy surface at different stages
of the motion are illustrated in : (a)$\theta = 0$; (b) $\theta = \frac{\pi}{2}$ and
(c) $\theta = \pi$. (bo) and (co) are just given as construction intermediates ; they show how the angle of pyramidalization before
rotation must invert algebraically for a rotation of 180°.

obtain the 5-dimensional (α, θ_1, β_1, θ_2, β_2) A-matrix. Second we reduce
it to the 3-dimensional (α, θ_1, θ_2)   A-matrix which is adiabaticity
corrected for pyramidalization of the terminal methylenes. The explicit
results are given in the next section. The probelm is actually very intricate. However, although the calculations are tedious, they are rather
straightforward.

2.4. Derivation of the Constraint-Matrices

    To obtain the constraint-matrices within the various models below,
we first settle the cartesian coordinates of the nuclei as functions of
α, θ_1, β_1, θ_2 and β_2. The molecule and the notation are shown in Figure
11. For given θ_i and β_i the position of the ith terminal group is constructed as follows : (i) the algebraic angle of pyramidalization β_i is
reported before the rotation takes place ($\theta_i = 0$, see Figure 11a, b_0 and
c_0), then (ii) the group is rotated of θ_i around the adjacent carbon-
carbon bond (see Figure 11a, b and c). The cartesian coordinates are
expressed in a frame where the center of gravity of the molecule is fixed
at the origin. The substituent groups on C^1 and C^2 (the carbon atoms

between which the bond breaks) are the same (mass m_1) but those on C^3 may be different (m_2). The rigid group $C^3H^{31}H^{32}$ ($M+2m_2$) is treated as an atom.

$$
\begin{matrix}
C^1 \\
(M)
\end{matrix}
\left|
\begin{aligned}
x_1 &= (1-2\chi)L\cos\alpha - \xi\lambda(a_1+a_2) \\
y_1 &= L\sin\alpha - \xi\lambda(b_1-b_2) \\
z_1 &= -\xi\lambda\,(c_1-c_2)
\end{aligned}
\right.
$$

$$
\begin{matrix}
C^2 \\
(M)
\end{matrix}
\left|
\begin{aligned}
x_2 &= (1-2\chi)L\cos\alpha - \xi\lambda(a_1+a_2) \\
y_2 &= -L\sin\alpha - \xi\lambda(b_1-b_2) \\
z_2 &= -\xi\lambda\,(c_1-c_2)
\end{aligned}
\right.
$$

$$
\begin{matrix}
C^3H^{31}H^{32} \\
(M+2m_2)
\end{matrix}
\left|
\begin{aligned}
x_3 &= -2\chi L\cos\alpha - \xi\lambda(a_1+a_2) \\
y_3 &= -\xi\lambda\,(b_1-b_2) \\
z_3 &= -\xi\lambda\,(c_1-c_2)
\end{aligned}
\right.
\qquad (2\text{-}8)
$$

$$
\begin{matrix}
H^{11(12)} \\
(m_1)
\end{matrix}
\left|
\begin{aligned}
x_{11(12)} &= (1-2\chi)L\cos\alpha + (1-\xi)\lambda a_1 - \xi\lambda a_2 + (-)\mu\sin\alpha\sin\theta_1 \\
y_{11(12)} &= L\sin\alpha + (1-\xi)\lambda b_1 + \xi\lambda b_2 - (+)\mu\cos\alpha\sin\theta_1 \\
z_{11(12)} &= (1-\xi)\lambda c_1 + \xi\lambda c_2 + (-)\mu\cos\theta_1
\end{aligned}
\right.
$$

$$
\begin{matrix}
H^{21(22)} \\
(m_1)
\end{matrix}
\left|
\begin{aligned}
x_{21(22)} &= (1-2\chi)L\cos\alpha - \xi\lambda a_1 + (1-\xi)\lambda a_2 - (+)\mu\sin\alpha\sin\theta_2 \\
y_{21(22)} &= -L\sin\alpha - \xi\lambda b_1 - (1-\xi)b_2 - (+)\mu\cos\alpha\sin\theta_2 \\
z_{21(22)} &= -\xi\lambda c_1 - (1-\xi)\lambda c_2 + (-)\mu\cos\theta_2
\end{aligned}
\right.
$$

where L is the CC bond length, λ and μ are respectively the cosine and sine projections of the CH bond length within CH_2, $\chi = \dfrac{M+2m_1}{\mathcal{H}}$, $\xi = \dfrac{2m_1}{\mathcal{H}}$ and $\mathcal{H}$ is the total mass : $\mathcal{H} = 3M+4m_1+2m_2$. a_i, b_i and c_i ($i=1,2$) are short notations for the auxiliary functions $a(\alpha, \beta_i, \theta_i)$, $b(\alpha, \beta_i, \theta_i)$ and $c(\beta_i, \theta_i)$, where :

$$
\begin{aligned}
a(\alpha, \beta, \theta) &= \cos\alpha\cos\beta + \sin\alpha\sin\beta\cos\theta \\
b(\alpha, \beta, \theta) &= \sin\alpha\cos\beta - \cos\alpha\sin\beta\cos\theta \\
c(\beta, \theta) &= -\sin\beta\sin\theta
\end{aligned}
$$

<u>Model in which α, θ_1, β_1, θ_2 and β_2 are five independent coordinates.</u>
The kinetic energy is :

$$
\begin{aligned}
T = A_{\alpha\alpha}\,\dot\alpha^2/2 + \sum_{\substack{i=1 \\ (j\neq i)}}^{2} \{ &A_{\theta_i\theta_i}\,\dot\theta_i^2/2 + A_{\alpha\theta_i}\,\dot\alpha\ddot\theta_i + A_{\beta_i\beta_i}\,\dot\beta_i^2/2 + A_{\alpha\beta_i}\,\ddot\alpha\dot\beta_i \\
&+ A_{\theta_i\beta_i}\,\dot\theta_i\dot\beta_i + A_{\theta_i\beta_j}\,\dot\theta_i\dot\beta_j\} + A_{\theta_1\theta_2}\,\dot\theta_1\dot\theta_2 + A_{\beta_1\beta_2}\,\dot\beta_1\dot\beta_2 \qquad (2\text{-}9)
\end{aligned}
$$

The elements of the matrix A are (see eq.0-1) :

$$A_{\alpha\alpha} = I_\lambda \ \{(1-\xi) \ (2-\sin^2\theta_1\sin^2\beta_1-\sin^2\theta_2\sin^2\beta_2)$$
$$+ 2\xi[\cos2\alpha(\cos\beta_1\cos\beta_2-\sin\beta_1\sin\beta_2\cos\theta_1\cos\theta_2)+\sin2\alpha(\sin\beta_1\cos\beta_2\cos\theta_1$$
$$+\cos\beta_1\sin\beta_2\cos\theta_2)]\}+ I_\mu \ (\sin^2\theta_1+\sin^2\theta_2) + I_L \ F(\alpha)$$
$$+ 2I_{L\lambda}[F(\alpha)(\cos\beta_1+\cos\beta_2)+\chi\sin2\alpha(\sin\beta_1\cos\theta_1+\sin\beta_2\cos\theta_2)]$$

$$A_{\theta_i\theta_i} = I_\lambda \ (1-\xi) \ \sin^2\beta_i + I_\mu \qquad\qquad (i = 1,2)$$

$$A_{\alpha\theta_i} = I_\lambda \ \sin\theta_i\sin\beta_i[(1-\xi)\cos\beta_i+\xi(\cos2\alpha\cos\beta_j+\sin2\alpha\sin\beta_j\cos\theta_j)]$$
$$+ I_{L\lambda} \ F(\alpha)\sin\theta_i \ \sin\beta_i \qquad\qquad (i = 1,2 \ ; \ j \neq i)$$

$$A_{\theta_1\theta_2} = I_\lambda \ \xi \ \sin\beta_1\sin\beta_2 \ (\cos2\alpha\sin\theta_1 \ \sin\theta_2 + \cos\theta_1 \ \cos\theta_2)$$

$$A_{\beta_i\beta_i} = I_\lambda \ (1-\xi) \qquad\qquad\qquad (i = 1,2) \qquad\qquad\qquad (2-10)$$

$$A_{\alpha\beta_i} = -I_\lambda \ \{(1-\xi)\cos\theta_i+\xi[\sin2\alpha(\sin\beta_i\cos\beta_j+\cos\beta_i\sin\beta_j\cos\theta_1\cos\theta_2)$$
$$+ \cos \alpha(\cos\theta_i\cos\beta_1\cos\beta_2-\cos\theta_j\sin\beta_1\sin\beta_2)]\}$$
$$- I_{L\lambda}[F(\alpha)\cos\theta_i\cos\beta_i+\chi\sin2\alpha\sin\beta_i] \quad (i = 1,2 \ ; \ j \neq i)$$

$$A_{\beta_1\beta_2} = I_\lambda \ \xi[\sin\theta_1\sin\theta_2\cos\beta_1\cos\beta_2-\cos2\alpha(\sin\beta_1\sin\beta_2-\cos\beta_1\cos\beta_2\cos\theta_1\cos\theta_2)$$
$$+ \sin2\alpha(\sin\beta_1\cos\beta_2\cos\theta_2+\sin\beta_2\cos\beta_1\cos\theta_1)]$$

$$A_{\theta_i\beta_i} = 0 \qquad\qquad\qquad\qquad (i = 1,2)$$

$$A_{\theta_i\beta_j} = I_\lambda \ \xi\sin\beta_i[\cos\theta_i\sin\theta_j\cos\beta_j-\sin\theta_i(\cos2\alpha\cos\theta_j\cos\beta_j$$
$$- \sin2\alpha\sin\beta_j)] \qquad\qquad (i,j = 1,2 \ ; \ i \neq j)$$

where $I_L = 2(M+2m_1)L^2$, $I_{L\lambda} = 2m_1L\lambda$, $I_\lambda = 2m_1\lambda^2$ and $I_\mu = 2m_1\mu^2$ are moments of inertia and $F(\alpha) = 1 - 2\chi \sin^2\alpha$.

Clearly, dynamical trajectories within this model are technically attainable. Unfortunately the 5-dimensional potential $V(\alpha,\theta_1,\beta_1,\theta_2,\beta_2)$ is lacking and so will remain. Indeed it represents an overwhelming task for little benefit from a static point of view.

Model in which β_1 and β_2 are secondary coordinates varying adiabatically with α, θ_1 and θ_2. The simple realtionship :

$$\beta_i = \beta_o(\alpha) \cos \theta_i \qquad\qquad (i = 1,2) \qquad\qquad (2-11)$$

is chosen because it fulfils all the requirements $\beta_i = \beta_o$ at $\theta_i = 0$, $\beta_i = 0$ at $\theta_i = 90°$ and $\beta_i = -\beta_o$ at $\theta_i = 180°$, ant it is a simple differentiable function whose derivative with respect to θ_i is continuous.

Within this model, the kinetic energy is :

$$T = \overline{A}_{\alpha\alpha} \dot{\alpha}^2/2 + \sum_{i=1}^{2}\{\overline{A}_{\theta_i\theta_i}\dot{\theta}_i^2/2 + \overline{A}_{\alpha\theta_i}\dot{\alpha}\dot{\theta}_i\} + \overline{A}_{\theta_1\theta_2}\dot{\theta}_1\dot{\theta}_2 \qquad (2\text{-}12)$$

The elements of the adiabaticity corrected matrix $\overline{A}$ are, from eq. 2-4 :

$$\overline{A}_{\alpha\alpha} = I_\lambda \{(1-\xi)(H_1+H_2)+2\xi[\cos 2\alpha(E_1 E_2 - F_1 F_2)+\sin 2\alpha(E_1 F_2 + E_2 F_1)+G_1 G_2]\}$$
$$+ I_\mu (\sin^2\theta_1+\sin^2\theta_2) + I_L F(\alpha)$$
$$+ 2I_{L\lambda}[F(\alpha)(E_1+E_2) + \chi\sin 2\alpha(F_1+F_2)]$$

$$\overline{A}_{\theta_i\theta_i} = I_\lambda (1-\xi) W_i + I_\mu \qquad\qquad (i = 1,2) \qquad (2\text{-}13)$$

$$\overline{A}_{\alpha\theta_i} = I_\lambda \{(1-\xi)(U_i F_i + V_i E_i + T_i G_i)+\xi[\cos 2\alpha(V_i E_j - U_i F_j)$$
$$+ \sin 2\alpha(U_i E_j + V_i F_j)+T_i G_j] + I_{L\lambda}[F(\alpha)V_i + \chi\sin 2\alpha\, U_i]$$
$$(i = 1,2 \; ; \; j \neq i)$$

$$\overline{A}_{\theta_1\theta_2} = I_\lambda \xi[\cos 2\alpha(V_1 V_2 - U_1 U_2) + \sin 2\alpha(U_1 V_2 + U_2 V_1) + T_1 T_2]$$

where several notations are the same as above but new auxiliary functions are defined :

$$E_i = \cos\beta_i (1 - \frac{d\beta_o}{d\alpha}\cos^2\theta_i)$$

$$F_i = (1 - \frac{d\beta_o}{d\alpha})\sin\beta_i \cos\theta_i$$

$$G_i = - \frac{d\beta_o}{d\alpha}\cos\beta_i \cos\theta_i \sin\theta_i$$

$$H_i = \cos^2\beta_i \sin^2\theta_i + (1 - \frac{d\beta_o}{d\alpha})^2 \cos^2\theta_i = E_i^2 + F_i^2 + G_i^2 \qquad (2\text{-}14)$$

$$T_i = \beta_o \cos\beta_i \sin^2\theta_i - \sin\beta_i \cos\theta_i$$

$$U_i = \beta_o \sin\beta_i \sin\theta_i \qquad\qquad (i=1,2)$$

$$V_i = \sin\theta_i (\beta_o \cos\beta_i \cos\theta_i + \sin\beta_i)$$

$$W_i = \beta_o^2 \sin^2\theta_i + \sin^2\beta_i = T_i^2 + U_i^2 + V_i^2$$

The same potential is used as in ref. [6b]. $\beta_o(\alpha)$ denotes, for given α, the symmetric pyramidalization of the terminal groups within the face-to-face conformation. Basic paths (zero excess energy) were obtained in ref. [6a] (see figures 12 and 14) for the ring angle motion and the symmetric pyramidalization. For cyclopropane this path (figure 12) is close to the reaction path (figure 13) ; thus the latter is kept as $\beta_o(\alpha)$. For tetramethylcyclopropane the basic path is close to a straight line joining the opened diradical to the reclosed ring (see Figure 14). Thus a linear function of α between these two boundaries is retained as $\beta_o(\alpha)$. The equations of motion are derived within the Lagrangian framework (eq. 0-4 and 0-5) by means of an inversion of $\overline{A}$ and partial

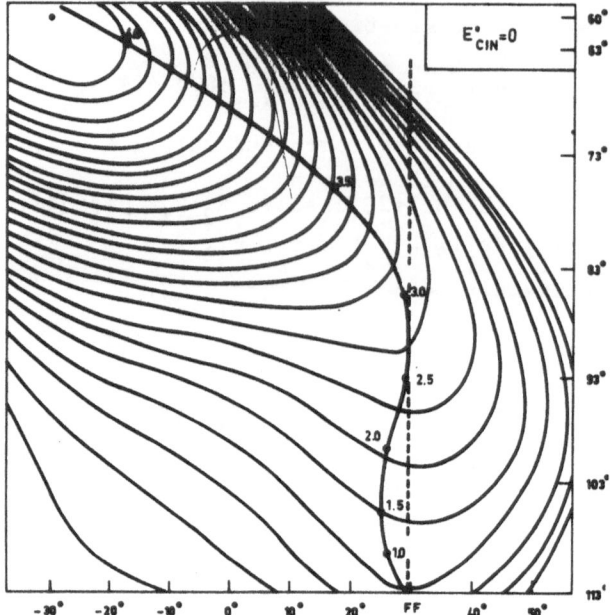

Figure 12 - The basic trajectory for the ring closure of
unsubstituted cyclopropane with zero kinetic energy in the
opened-ring conformation FF.

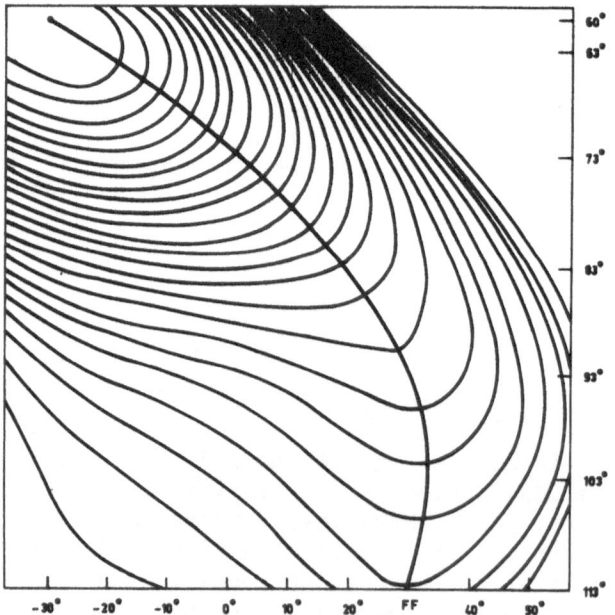

Figure 13 - Static reaction path for the ring closure from FF
to cyclopropane. The abcissas are the values of the angle of
symmetric pyramidalization, the ordinates the values of the
angle 2α. The energy gap between two isoenergetic lines is
2.5 kcal/mol.

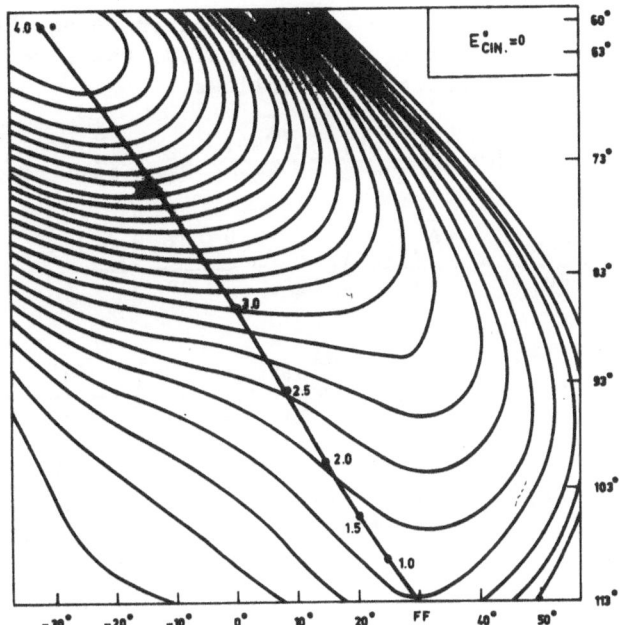

Figure 14 – The basic trajectory for the ring closure of 1, 1, 2, 2-tetramethyl-cyclopropane with zero kinetic energy in the opened ring conformation FF.

differentiation of its elements with respect to α, θ_1 and θ_2 respectively. The integrator is a simple fourth-order Runge-Kutta[20].

Model in which the terminal groups are trigonal throughout the motion. The kinetic energy is now diagonal :

$$T = (a_{\alpha\alpha}\,\dot{\alpha}^2 + \sum_{i=1}^{2} a_{\theta_i\theta_i}\,\dot{\theta}_i^2)/2 \qquad (2\text{-}15)$$

The a's are derivable from either $A_{\alpha\alpha}$ and $A_{\theta_i\theta_i}$ (eqs.2-10) or $\overline{A}_{\alpha\alpha}$ and $A_{\theta_i\theta_j}$ (eqs.2-13) by replacing everywhere β_1 and β_2 in eqs.2-10 and β_1, β_2, β_0 and $\dfrac{d\beta_0}{d\alpha}$ in eqs.2-13 by 0. This results in :

$$a_{\alpha\alpha} = 2I_\lambda\,(1-2\xi\sin^2\alpha) + I_\mu\,(\sin^2\theta_1+\sin^2\theta_2) + (I_L +4I_{L\lambda})F(\alpha)$$
$$a_{\theta_i\theta_i} = I_\mu \qquad (i = 1,2) \qquad (2\text{-}16)$$

These expressions can be readily identified with eqs. 11 to 15 in ref. [6b]. This model is used in ref. [6b] and in section 1 of the present article.

2.5. Results and discussion

The results presented below all are obtained within the pyramidalization adiabaticity corrected (PAC) model . They are compared with results obtained previously within the simplified trigonal (ST) model .

The expression of the kinetic energy is given in eq. 2-12 and the
A-matrix elements are given in eq. 2-13. Finally, the three Lagrangian
equations of motion are established as follows :

$$S_\alpha = \frac{1}{2} \frac{\partial \overline{A}_{\alpha\alpha}}{\partial \alpha} \dot{\alpha}^2 + \sum_{i=1}^{2} [(-\frac{1}{2} \frac{\partial \overline{A}_{\theta_i\theta_i}}{\partial \alpha} + \frac{\partial \overline{A}_{\alpha\theta_i}}{\partial \theta_i}) \dot{\theta}_i^2 + \frac{\partial \overline{A}_{\alpha\alpha}}{\partial \theta_i} \dot{\alpha}\dot{\theta}_i]$$

$$+ (\frac{\partial \overline{A}_{\alpha\theta_1}}{\partial \theta_2} + \frac{\partial \overline{A}_{\alpha\theta_2}}{\partial \theta_1} - \frac{\partial \overline{A}_{\theta_1\theta_2}}{\partial \alpha}) \dot{\theta}_1\dot{\theta}_2 + \frac{\partial \mathcal{V}(\alpha,\theta_1\theta_2)}{\partial \alpha}$$

$$\hspace{10cm} (2\text{-}17)$$

$$S_{\theta_i} = (-\frac{1}{2} \frac{\partial \overline{A}_{\alpha\alpha}}{\partial \theta_i} + \frac{\partial \overline{A}_{\alpha\theta_i}}{\partial \alpha}) \dot{\alpha}^2 + \frac{1}{2} \frac{\partial \overline{A}_{\theta_i\theta_i}}{\partial \theta_i} \dot{\theta}_i^2 + (-\frac{1}{2} \frac{\partial \overline{A}_{\theta_j\theta_j}}{\partial \theta_i} + \frac{\partial \overline{A}_{\theta_i\theta_j}}{\partial \theta_j}) \dot{\theta}_j^2$$

$$+ \frac{\partial \overline{A}_{\theta_i\theta_i}}{\partial \alpha} \dot{\alpha}\dot{\theta}_i + (-\frac{\partial \overline{A}_{\alpha\theta_j}}{\partial \theta_i} + \frac{\partial \overline{A}_{\alpha\theta_i}}{\partial \theta_j} + \frac{\partial \overline{A}_{\theta_i\theta_j}}{\partial \alpha}) \dot{\alpha}\dot{\theta}_j + \frac{\partial \overline{A}_{\theta_i\theta_i}}{\partial \theta_j} \dot{\theta}_i\dot{\theta}_j$$

$$+ \frac{\partial \mathcal{V}(\alpha,\theta_1,\theta_2)}{\partial \theta_i} \hspace{3cm} (i=1,2 \; ; \; j\neq i)$$

and, in matrix notation :
$$\begin{bmatrix} \ddot{\alpha} \\ \ddot{\theta}_1 \\ \ddot{\theta}_2 \end{bmatrix} = -\overline{A}^{-1} \begin{bmatrix} S_\alpha \\ S_{\theta_1} \\ S_{\theta_2} \end{bmatrix} \hspace{3cm} (2\text{-}18)$$

The potential $\mathcal{V}(\alpha,\theta_1,\theta_2)$ is the same as in ref.[6b]. It takes into
account correctly the adiabatic pyramidalization motion of the terminal
groups. The numerical integration of the three coupled second-order
differential eqs. 2-18 requires six initial conditions. The three
values $2\alpha°$, $\theta_1°$ and $\theta_2°$ determine the molecular geometry at the starting
point (all the trajectories start with the cyclopropane in its equi-
librium geometry, i.e. $2\alpha° = 60°$, $\theta_1° = \theta_2° = 0$). In addition we specify
the total internal energy in the molecule (E_{tot}), the fraction of it
attributed to the "rotations" at the starting point ($E_{rot}°$) and the
manner in which $E_{rot}°$ is distributed among the two rotors ($\delta°$ such that
tg $\delta° = \dot{\theta}_1°/\dot{\theta}_2°$ where the $\dot{\theta}°$'s are the initial rotational velocities).
Then the relationship :

$$\dot{\theta}_2° = [2E_{rot}°/(\overline{A}_{\theta_1\theta_1}° tg^2\delta° + 2\overline{A}_{\theta_1\theta_2}° tg\delta° + \overline{A}_{\theta_2\theta_2}°)]^{1/2} \hspace{2cm} (2\text{-}19)$$

is used. Finally the relationship :

$$\dot{\alpha}° = -B° \pm \{(B°)^2 + 2[E_{tot} - E_{rot}° - \mathcal{V}(\alpha°,\theta_1°,\theta_2°)]/\overline{A}_{\alpha\alpha}°\}^{1/2} \hspace{1cm} (2\text{-}20)$$

where $B° = (\sum_{i=1}^{2} \overline{A}_{\alpha\theta_i} \dot{\theta}_i°)/\overline{A}_{\alpha\alpha}°$ defines the initial angular velocity of the ring angle.

About five hundred trajectories were studied. The major conclusions obtained in the previous simplified studies (ref. [6b] and section 1 above) are conserved, namely (i) all the dynamical reactive trajectories are closely connected to the reaction path emerging from the static calculations ; (ii) the initial terminal groups "rotational" (actually vibrational) energy required for a trajectory to be reactive is much greater than expected from the static calculations ; (iii) distinct "reactive bands" appear when varying the initial "rotational" energy (band structure) ; (iv) for all the reactive trajectories, a single rotation of 180° of the terminal groups occurs.

Some noticeable modifications of the results in ref. [6b] are illustrated in Tables II and III. First, the reactive bands are slightly wider,

$E^°_{rot}$ kcal/mol $\delta°$ deg	6	8	10	12	14	16	18	20	22	24	26	28	30
− 45						EE_D	EE_D	EE_D	EE_D	EE_D	EE_D	(EE_D) **	
						EE_D	EE_D	EE_D	EE_D	EE_D	EE_D		
0					(EF+EF) *	EF	EF	EF	EF	EF	EF		
						EF	EF	EF	EF	EF			
+ 45			EE_C	EE_C	EE_C	EE_C	EE_C	EE_C	EE_C	EE_C	EE_C		
			EE_C	EE_C	EE_C		EE_C	EE_C		EE_C			

Table II — Nature of the half-way point for isomerization of cyclopropane depending on the values of E_{rot} at $\delta° = 45°$ (pure conrotatory motion), $\delta° = 0$ (dominant single rotation) and $\delta° = -45°$ (pure disrotatory motion), when $E_{tot} = 63$ kcal/mol. The upper terms are for the PAC model and the lower terms are for the ST model [6b]. The blank terms of the array correspond to non reactive trajectories. EE_C and EE_D mean that the reaction intermediate is in the edge-to-edge conformation and is reached via, respectively, a conrotatory and a disrotatory motion ; the reaction product is the optical isomer. EF means that the reaction intermediate is in the edge-to-face conformation and the reaction product is the geometrical isomer. * denotes a trajectory in which the rotation of one of the two terminal group takes place and, before the geometrical isomer is formed, the rotation of the other terminal group occurs. Thus the reaction product is the optical isomer formed via two successive EF intermediates. ** denotes a disrotatory trajectory in which the EE_D intermediate is formed and re-formed several times back and forth, i.e. the representative point of the reaction is trapped in the small well surrounding the half-way point (rotational angles between 60° and 120°).

$E°_{rot}$ (kcal/mol) / $δ°$, deg	4	6	8	10	12	14	16	18	20	22	24	26	28
- 45				EE_D	EE_D	EE_D	EE_D	EE_D	EE_D	EE_D	EE_D		
					EE_D	EE_D	EE_D	EE_D	EE_D	EE_D	EE_D	EE_D	
0			$*$ (EF+EE)/EE	EF	EF	EF	EF	EF	EF	EF	EF		
						EF	EF	EF	EF	EF	EF		
+ 45	EE_C	EE_C	EE_C	EE_C	EE_C	EE_C	EE_C	EE_C	EE_C	EE_C	EE_C	EE_C	
					EE_C	EE_C	EE_C	EE_C	EE_C	EE_C	EE_C	EE_C	

Table III - Nature of the half-way point for isomerization of 1,1,2,2-tetramethylcyclopropane, depending on the values of $E°_{rot}$ at $δ°$ = -45°, 0 and +45° when E_{tot} = 67 kcal/mol. The notations are the same as in Table II. * denotes a trajectory in which one of the two terminal groups rotates of 90° (EF conformation) then stops near the half-way point so that the other group has time to rotate and the EE intermediate is formed whereas a single group is rotating. The reaction product is the optical isomer.

in particular in the vicinity of $δ°$ = 0 (single rotation motion), in the PAC case than in the ST case. Second on the sides of the reactive bands there appear, in the PAC case, some very intricate trajectories (indicated by stars in Tables II and III) in which the representative point hesitates about where to go in the region of the half-way point. In some trajectories (for example see Table II for $δ°$ = -45° and $E°_{rot}$ = 28 kcal/mol) the representative point is trapped in this region and never goes out of it. This certainly results in a modification of the entropy factor of Benson [12]. However, on the average, the durations of the various isomerization processes are the same in the PAC case as in the ST case.

It should be emphasized that the PAC model cannot be used to determine improved values of the total energy thresholds for reactive trajectories to compare with the values of the thresholds reported in Table I. Indeed in the threshold region the adiabatic pyramidalization motion acts as an energy reservoir which can receive energy from the rotational motion when this is hindered and stocks this energy until the passage over the transition state is possible to the representative point of the reaction. Thus reactive trajectories appear at much lower total energy in the PAC case than in the ST case. Clearly these low energy tails of the reactive bands are physically irrelevant. However at higher total energy the pyramidalization acts only as a perturbative correction to the rotational motion and is correctly accounted for in the PAC model.

CONCLUSION

This communication accounts for the present state of the Theory of Constrained Systems and its application to organic chemistry dynamical problems. In part 1 the introduction of substituent effects forms most probably the end of the long dynamical study of the isomerizations of cyclopropane. In part 2 an extension of the Theory of Constrained Systems is introduced. It aims at making identical the models used in the static quantum mechanical calculations of the potential on the one hand and in the dynamical calculations of trajectories on the other hand, including the static optimization parameters. It results in an improvement of the theory but requires to be handled carefully.

NOTES

1 – We extend here to classical mechanics the usual quantum-mechanical Dirac's notation.

2 – It is possible to obtain reactive trajectories at lower total energy by relaxing all the arbitrary constraints in the starting molecule, i.e. $2\alpha^\circ=60^\circ$, $\theta_1^\circ=\theta_2^\circ=0$ and $\alpha_0>0$. Such minimal total energy trajectories were indeed obtained by starting from half-way points ($2\alpha_0=113^\circ$, $\theta^\circ=90^\circ$ along with either $\theta_1^\circ=\theta_2^\circ=\theta^\circ$, or $\theta_1^\circ=-\theta_2^\circ=\theta^\circ$ or $\theta_1^\circ=\theta^\circ$ and $\theta_2^\circ=0$) and integrating backwards until the ring was reclosed. Nevertheless the purpose of the present study is to compare the dynamical results for both unsubstituted and substituted cyclopropanes within the same constraints. A complete study of total energy minimalized trajectories is beyond the scope of this worK

3 – The transition state for geometrical isomerization lies at a rotation angle of 90°. This very unfavorable situation is partly cancelled by the fact that the rotation of a single group is, for a given amount of excess total energy, noticeably faster than the concerted rotation of both groups. Indeed the moment of inertia associated with the rotation of a single group is half that for concerted rotation of both groups. This explains why the total energy threshold for geometrical isomerization is not higher than that for optical isomerization by conrotatory pathway.

4 – A similar dynamical effect was first discussed by Porter et al, in the study of the reactive collision system $H_2 + I_2 \rightarrow 2HI$, for which a pathway other than the energetically favoured one is dominant.

5 – When $E_{tot} \leq 67$ kcal/mole, the N=1 type band is not observed whenever $\delta^\circ \leq -5^\circ$. Thus in particular it does not appear for the synchronous disrotatory motion.

REFERENCES

1. X. Chapuisat and Y. Jean, Topics Current Chem. 68 (1976)
2. D.L. Bunker and M.D. Pattengill, J. Chem. Phys. 53, 3041 (1970)
   D.L. Bunker, J. Chem. Phys. 57, 332 (1972)
   D.L. Bunker and W.L. Hase, J. Chem. Phys. 59, 4621 (1973)
   H.H. Harris and D.L. Bunker, Chem. Phys. Lett. 11, 433 (1971)
3. D.L. Bunker and N.C. Blais, J. Chem. Phys. 37, 2713 (1962)
   N.C. Blais and D.L. Bunker, J. Chem. Phys. 39, 315 (1963)
   D.L. Bunker and N.C. Blais, J. Chem. Phys. 41, 2377 (1964)
   M. Karplus and L.M. Raff, J. Chem. Phys. 41, 1267 (1964)
   L.M. Raff, J. Chem. Phys. 44, 1202 (1966)
   L.M. Raff and M. Karplus, J. Chem. Phys. 44, 1212 (1966)
4. D.L. Bunker and M.D. Pattengill, Chem. Phys. Lett. 4, 315 (1969)
   T. Válencich and D.L. Bunker, Chem. Phys. Lett. 20, 50 (1973)
5. I.S. Wang and M. Karplus, J. Amer. Chem. Soc. 95, 8160 (1973)
6. a) Y. Jean and X. Chapuisat, J. Amer. Chem. Soc. 96, 6911 (1974)
   b) X. Chapuisat and Y. Jean, J. Amer. Chem. Soc. 97, 6325 (1975)
7. E.A. McCullough and R.E. Wyatt, J. Chem. Phys. 54, 3578 (1971)
   E.A. McCullough and R.E. Wyatt, J. Chem. Phys. 54, 3592 (1971)
   J.C. Light, Met. Comp. Phys. 10 (1971)
   W.H. Miller, Acc. Chem. Res. 4, 161 (1971)
   W.H. Miller, Adv. Chem. Phys. 25 (1974)
8. J.A. Berson, L.D. Pedersen and B.K. Carpenter, J. Amer. Chem. Soc. 98, 122 (1976)
9. R.G. Bergman, Free Radicals (ed. J.K. Kochi) Vol. I, chap. 5, Wiley (1973)
10. W.L. Carter and R.G. Bergman, J. Amer. Chem. Soc. 90, 7345 (1968)
    R.G. Bergman and W.L. Carter, J. Amer. Chem. Soc. 91, 7411 (1969)
11. J.A. Berson and J.M. Balquist, J. Amer. Chem. Soc. 90, 7343 (1968)
12. H.E. O'Neal and S.W. Benson, J. Phys. Chem. 72, 1866 (1968)
13. W. von E. Doering and K. Sachdev, J. Amer. Chem. Soc. 96, 1168 (1974)
    R.J. Crawford and T.R. Lynch, Can. J. Chem. 46, 1457 (1968)
14. M.C. Flowers and H.M. Frey, Proc. Roy. Soc. 257, 122 (1960)
    M.C. Flowers and H.M. Frey, Proc. Roy. Soc. 260, 424 (1961)
    D.W. Setzer and B.S. Rabinovitch, J. Amer. Chem. Soc. 86, 564 (1964)
    H.M. Frey and D.C. Marshall, J. Chem. Soc. P. 191 (1965)
    H.M. Frey, Adv. Phys. Org. Chem. 4, 147 (1966)
15. R.N. Porter, D.L. Thompson and L.B. Sims, J. Amer. Chem. Soc. 92, 3208 (1970)
16. X. Chapuisat, Y. Jean and L. Salem, Chem. Phys. Lett. 37, 119 (1976)
17. A. Gavezotti and M. Simonetta, private communication
18. R.D. Present, Kinetic Theory of Gases, McGraw-Hill (1958)
    S.W. Benson, The Foundations of Chemical Kinetics, McGraw-Hill (1960)
    P.J. Robinson and K.A. Holbrook, Unimolecular Reactions, Wiley (1972)
    W. Forst, Theory of Unimolecular Reactions, Academic Press (1973)
19. Y. Jean, L. Salem, J.S. Wright, J.A. Horsley, C. Moser and R.M. Stevens, Pure Appl. Chem. Suppl. 1, 197 (1971)
    J.A. Horsley, Y. Jean, C. Moser, L. Salem, R.M. Stevens and J.S. Wright, J. Amer. Chem. Soc. 94, 279 (1972)
    Y. Jean, Thesis, Isomérisation Géométrique du Cyclopropane, Orsay (1973)
20. C.W. Gear, Numerical Initial Value Problems in Ordinary Differential

QUESTIONS ET COMMENTAIRES CONSECUTIFS A L'EXPOSE DE X. CHAPUISAT

R. Lefebvre : Des effets de translation et rotation en bloc de la molécule ne contaminent-ils pas les effets dynamiques que vous étudiez ?

Réponse : La translation en bloc est soigneusement écartée. Pour cela
nous utilisons un référentiel où le centre de gravité de la molécule
est fixé à l'origine.Ceci complique d'ailleurs les équations du mouvement mais s'avère nécessaire. Pour ce qui est de la rotation en bloc,
la séparer totalement (en imposant la nullité du moment angulaire total)
s'avère irréalisable. Les équations du mouvement deviennent beaucoup
trop compliquées. Néanmoins, nous avons pris soin de placer géométrique-
ment la molécule dans le référentiel fixe de sorte qu'elle ne tourne pas
en bloc. Tout au plus, peut-elle effectuer de légères oscillations lors
de la rotation des groupes terminaux. Mais celles-ci sont si faibles que
leur influence dynamique peut certainement être négligée.

R. Lefebvre : Les surfaces de potentiel que vous utilisez pour le cyclopropane et la molécule substituée sont strictement les mêmes. Les différences existant entre les deux systèmes ne sont-elles pas en fait bien
supérieures à celles que vous discutez dans la comparaison des effets
dynamiques ?

Réponse : C'est en effet probable. Les principales différences de surface
entre le cyclopropane et la molécule tétrasubstituée sont les suivantes :
(1) le point de départ (2 $\alpha° = 60°$, $\theta_1^° = \theta_2^° = 0$) est plus haut par dé-
stabilisation stérique ; (2) pour la même raison, l'état de transition
est plus haut et (3) il est déporté, pour les mouvements conrotatoire
et disrotatoire, vers la valeur $\theta = 90°$. On pourrait modifier empiri-
quement la surface pour rendre compte des ces différences. Tel n'était
pas notre but. Notre but fut d'étudier les effets de masse, en compa-
rant qualitativement les comportements dynamiques des deux systèmes sur
une même surface, en particulier pour la phase de rotation des groupes
terminaux dans la vallée de potentiel supérieure. Nous n'avons ici aucune
prétention à calculer une énergie d'activation dynamique dans le cas
substitué. Par contre, les effets que nous mentionnons contribuent certainement à la variation du facteur entropique. Le fait que nous ayons choisi
le 1,1,2,2-tetraméthylcyclopropane nous permet seulement de bénéficier
de la symétrie élevée de cette molécule. Notre comparaison s'applique
aussi bien à n'importe quel cyclopropane lourdement substitué.

Dannenberg : Un commentaire. Simonetta a trouvé, pour les cyclopropanes
substitués, une inversion de l'ordre des énergies des états de transition.
Il n'a cependant pas calculé les surfaces complètes et la méthode de cal-
cul qu'il a utilisée semble être moins fine que celle utilisée par Salem
pour le cyclopropane. Une question maintenant. Dans quelle mesure a-t-on
le droit de choisir les coordonnées dynamiques ? - Ne doit-on pas utiliser des coordonnées normales ?

Réponse : Le choix des coordonnées n'est pas effectué par les "dynami-
ciens", mais par les quantochimistes qui calculent les surfaces de potentiel. Notre propos n'est que d'utiliser -et exactement si possible- les
modèles moléculaires développés par les quantochimistes. Il est certain
qu'au niveau de la dynamique, des ambiguïtés peuvent procéder de cette
attitude. En effet, certains degrés de liberté négligés, à juste titre,
statiquement peuvent avoir un rôle dynamique non négligeable, ne serait-
ce qu'à cause de leur multitude dans les grandes molécules. Ils peuvent
aussi jouer le rôle d'un réservoir d'énergie et agir par des effets
d'adiabaticité. Pour les réactions unimoléculaires, il est parfois
possible d'estimer leur influence au moyen de théories statistiques
(RRKM par exemple). Il vaut ici la peine de rappeler que le but -frag-
mentaire- de la dynamique chimique n'est pas de décrire le comportement
dynamique exact des molécules en réaction, mais seulement de mettre en
évidence certains effets typiques de ce comportement. Pour ce qui est
de l'usage de coordonnées normales, deux cas sont à distinguer : s'il
s'agit des coordonnées normales des réactants, celles-ci peuvent être
utiles pour déterminer la distribution des conditions initiales ; mais
elles ne sont pas utilisables comme coordonnées de réaction. S'il s'agit
des coordonnées normales adiabatiques obtenues en séparant en tout point
de la surface la coordonnée de réaction locale des coordonnées locales
non dissociatives, de telles coordonnées peuvent parfaitement être
utilisées en dynamique mais sont en général d'un usage difficile.

THEORETICAL STUDY OF SOME SIMPLE ORGANIC REACTIONS

Yves JEAN
Laboratoire de Chimie Organique Physique
Université Louis Pasteur
67008 STRASBOURG Cedex (France)

ABSTRACT

The mechanisms of four gas phase organic reactions are studied through ab-initio LCAO-MO-SCF calculations. These reactions are : (i) cis-trans isomerization of cyclopropane ; (ii) decomposition of cyclobutane into two ethylene molecules ; (iii) the Diels-Alder reaction ; (iv) decomposition of 1-pyrazolines into cyclopropane and nitrogen. The structure and the electronic properties of the various transition states are discussed. In particular, it is shown that a diradical can be either a transition state, a secondary intermediate, or a "transient point" on a hillside.

1. INTRODUCTION

The purpose of this article is to survey some work undertaken in the Laboratory of Theoretical Chemistry at Orsay, in the field of potential energy surface calculations. These calculations have been carried out on simple gas phase organic reactions, which are likely to proceed on a single potential surface connecting the reactant and product ground states. These reactions are :

1) Cis-trans isomerization of cyclopropane[1];
2) Decomposition of cyclobutane into two ethylene molecules[2];
3) The Diels-Alder reaction[3];
4) Decomposition of 1-pyrazolines into cyclopropane and nitrogen[4].

For reactions 1 and 2, the main problem prior to calculations, was the possible existence of a secondary minimum along the optimum reaction path. Such a minimum was strongly suggested by thermochemical calculations on tri- and tetramethylene diradicals, but had not been observed experimentally. In the case of reactions 3 and 4, a number of mechanisms had been proposed. Thus, our purpose was (i) to study typical reaction paths corresponding to these mechanisms, (ii) to analyse the structures and the electronic properties of the most accessible transition states.

53

R. Daudel, A. Pullman, L. Salem, and A. Veillard (eds.), Quantum Theory of Chemical Reactions, Volume I, 53-68.
Copyright © 1979 by D. Reidel Publishing Company.

The calculations were performed by the following approach : a) optimization of reactant and product geometries ; b) scan of typical
pathways using a minimal basis set ; c) optimization of the transition
states at this level ; d) recomputation of the energy pathways through
the previously optimized transition states, with either a minimal
(reactions 1 and 2) or, an extended (reactions 3 and 4), basis set.

The series of GAUSSIAN 70 computer programs[5] were employed, except
in the case of reaction 1[6], and the minimal STO-3G[7] and extended 4-31G[8]
basis sets were used. Two calculations were performed concurrently.
First, a "closed-shell" calculation using the classic Hartree-Fock
operator for a closed-shell configuration.

$$F = h + \sum_{j<a} (2J_j - K_j) + (2J_a - K_a)$$

where the usual notation is employed and where the highest doubly
occupied and the lowest unoccupied orbitals are labelled, respectively
a and b. The companion "open-shell" claculation uses the single restric-
ted open-shell Hamiltonian suggested by Nesbet[9].

$$F = h + \sum_{j<a} (2J_j - K_j) + (J_a - 1/2\ K_a) + (J_b - 1/2\ K_b)$$

in which both the orbitals a and b are assumed to be singly occupied. A
simple SCF calculation using either Hamiltonian, with the direct computation of the diagonal energies of the singlet configuration $^1a^2$ or
^1ab, is inappropriate for describing loosely bound transition states
or diradicals[10]. The high-energy ionic terms in the molecular orbital
description of a singlet diradical can be eliminated via a 3x3 configuration treatment[10]. The three configurations to be mixed are $^1a^2$, ^1ab
and $^1b^2$. For each point of the surfaces, both closed- and open-shell
calculations are performed, and we accept the calculation which gives
the lowest final energy after completion of the CI. Frequently, it is
found that the molecular orbitals produced by the Nesbet calculation
are preferable everywhere on the surfaces except in the neighbourhood
of reactants and products.

2.   CIS-TRANS ISOMERIZATION OF CYCLOPROPANE

This reaction is formally one of the simplest in organic chemistry,
involving only rotation by 180° of a methylene group. In the year of
its discovery[11] (1958), two mechanisms were proposed (figure 1) for the
isomerization. In the first mechanism, a methylene group rotates without
deformation of the carbon ring[12]. The second involves the intermediacy
of a trimethylene diradical formed by the cleavage of a carbon-carbon
bond[11]. Rotation about a carbon-carbon bond in the diradical, and
recyclization give the isomerized cyclopropane. This latter mechanism
has been widely considered to be the most likely of the two proposed,
especially since the intermediate structure involved in the first

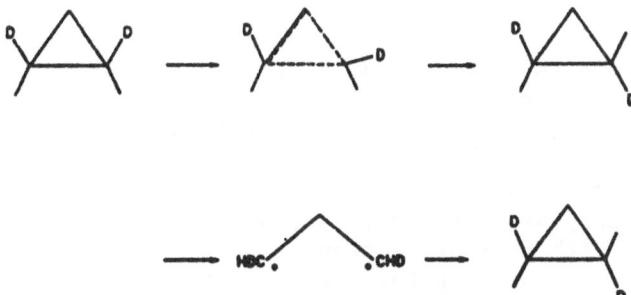

Figure 1 : Two limiting mechanisms for the cis-trans isomerization of cyclopropane : a) without deformation of the carbon
ring ; b) with initial breaking of one C-C bond and intermediate formation of the trimethylene diradical.

mechanism rather seems to be a precursor of the decomposition of cyclopropane into ethylene and methylene. Furthermore, a M.O. analysis of
this latter structure shows that it possesses a very high energy[13], and
it is a point reconfirmed in our calculations[1].

However, the problem of the properties of the trimethylene diradical remained unresolved. Thermochemical calculations by Benson, suggest
that this diradical is a rather secondary minimum on the potential
energy surface[14-16]. The principle of these calculations is indicated
in figure 2. The enthalpy of formation of the diradical from cyclopropane can be estimated from the heat to hydrogenate cyclopropane

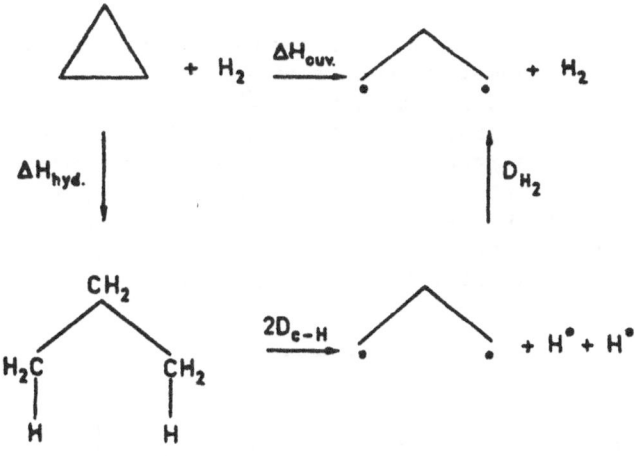

Figure 2 : Thermochemical calculation of the heat of formation
of the trimethylene diradical.

to propane (-37.8 kcal/mol), the heat of a C-H bond in propane (98 kcal/
mol) and the heat of recombination of two hydrogen atoms (-104 kcal/mol).
However, note that the assumption is made that the energy required to
break a C-H bond in propane and in the radical so formed, are equal.
The result of this calculation is ΔH_f = 54.2 kcal/mol whereas the activation energy is 64 kcal/mol. The conclusion drawn by Benson is that the
trimethylene diradical should be a rather stable secondary minimum (by
about 10 kcal/mol) along the reaction path for the cis-trans isomerization of cyclopropane. The results of our ab initio calculations are in
contradiction with this prediction[1], as no secondary minimum is found
along the minimum energy path. The transition state is a diradical in
which all residual bonding between carbons 1 and 3 is destroyed.

However, not only the transition state, but all the upper part of
the potential energy curve corresponds to diradical structures, of
differing conformations. Thus, quantum calculations have shown that the
energy of the trimethylene diradical is rather sensitive (6-8 kcal/mol)
to its conformation. On the other hand, thermochemical calculations
give a single value for the energy of the diradical, and this energy
might well be that of a diradical more stable than the transition state.
The absence of a barrier for the reclosure of the 1,3-diradicals has
been recently confirmed experimentally[17].

Finally, we have found that the easiest motion in the trimethylene
diradical is a synchronous, conrotatory rotation of the two terminal
groups. The kinetic analysis of racemization and trans-cis isomerization
in trans-1,2-dideuterocyclopropane[18] has confirmed that the synchronous
rotation is less energetic than the rotation of a single group.

3.   THERMAL DISSOCIATION OF CYCLOBUTANE INTO TWO ETHYLENE MOLECULES

Three main mechanisms can be invoked for this reaction ; a) a concerted 2s+2s process. However, such a process is symmetry forbidden
according to the Woodward-Hoffmann rules[19], and ab-initio calculations
for this path reveal a large activation energy[20]; b) a concerted,
symmetry allowed, 2s+2a process. Examination of this process, however,
reveals strong steric interactions, causing a high energy of activation.
Attempts to show that the reaction proceeds in such a concerted 2s+2a
fashion, have proven negative[21] ; c) a non-concerted process, leading
to the formation of the intermediate tetramethylene diradical (figure 3)

Figure 3 : Decomposition of cyclobutane through the tetramethylene diradical.

has been generally considered most likely.

The tetramethylene diradical has been predicted to be thermodynamically stable. The pyrolysis of cyclobutane requires an activation
energy of 62.5 kcal/mol[22]. Benson has estimated the enthalpy of formation of $.CH_2CH_2CH_2CH_2.$ to be 4 kcal/mol below the activation energy of
the reaction[23], and thus concludes that the diradical represents an
energy well on the potential energy surface of the reaction.

In theoretical study of this reaction, by G. Segal[2], all the geometrical parameters have been taken into account except the CH bond
distances[24]. Two minima were found for the tetramethylene diradical,
corresponding to gauche (G) and trans (T) structures (Figure 4), the
trans isomer being slightly more stable with an energy barrier between

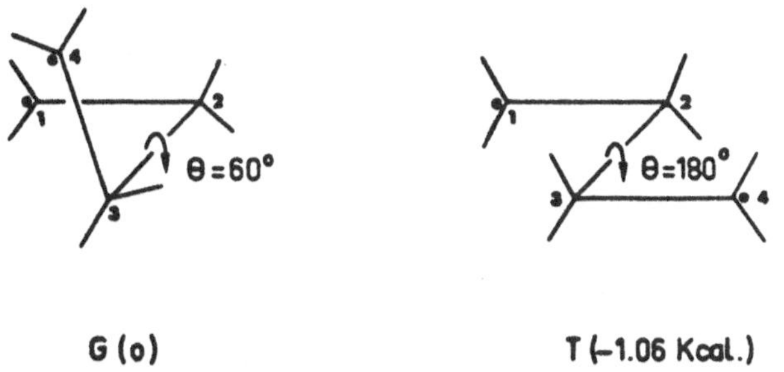

G (o)   **T (−1.06 Kcal.)**

Figure 4 : Gauche and trans structures of the tetramethylene
diradical

G and T of 3.6 kcal/mol.

G and T are minima with respect to dissociation into two ethylene
molecules (Figure 5). The energy barrier is 2.3 kcal/mol at C_2C_3=1.65 Å,
for T and 3.55 kcal/mol at 1.95 Å, for the G structure. The striking
difference between the values of C_2C_3 at the maxima of the potential
energy curves for the dissociation of G and T can be ascribed to residual C_1C_4 bonding. In the gauche conformation, as the C_2C_3 bond elon-
gates, the molecule tends to compensate for increasing energy by re-
establishing some C_1-C_4 bonding, made possible by an initial decrease
of the $C_1C_2C_3$ angle. This initial compensatory effect results in a
late barrier. However this compensation cannot take place in the dissociation from the trans minimum, and the barrier is thus calculated for
a shorter C_2C_3 distance. Finally, it is worth nothing that the dissociation from G and T require approximately the same activation energy,
respectively 3.55 and 3.60 kcal/mol (O=energy of G).

G and T are also minima with respect to the recyclization to cyclobutane. It is obvious for the trans form which must reclose via a
geometry in the vicinity of the gauche conformation ; the barrier for
the reclosure of the gauche form is estimated at 2.0 kcal/mol. Finally,
the energetic profile of the dissociation reaction through the gauche

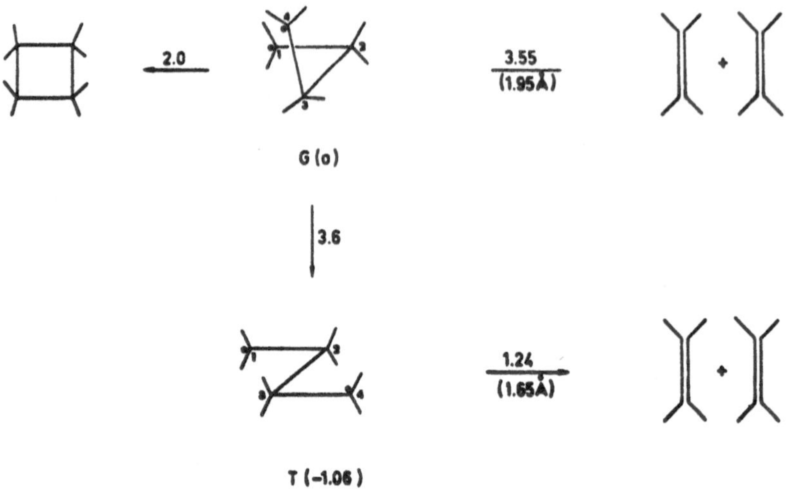

Figure 5 : Main features of the potential energy surface for the decomposition of cyclobutane into two ethylene molecules. The energies are in kcal/mol above G.

conformation shows a secondary minimum at the top of the pathway. The computed stability of this secondary minimum is very close to the thermochemical estimate. Thus, in this second thermal reaction proceeding through a diradical species, ab initio calculations and thermodynamic estimations are in full agreement.

4 .   THE DIELS-ALDER REACTION

      The Diels-Alder reaction is one of the most important in organic chemistry. A precise knowledge of the energy profile was necessary to further our understanding of the stereochemistry at both diene and dienophile, the endo-exo ratio, the regioselectivity, the influence of catalysts, of pressure and the "syn-anti" effect[25]. Salem and coworkers first considered two limiting mechanisms : (i) concerted, symmetry allowed, 4s+2s addition[19], which is characterized by the synchronous formation of the two new CC bonds ; (ii) non-concerted addition leading to cyclohexene via the hex-2-ene 1,6 diyl diradical. In this latter mechanism, the two new CC bonds form successively and the intermediate diradical should be a secondary minimum on the potential energy surface. However, although experimental data favour the concerted mechanism[26], MINDO/3 calculations by Dewar and coworkers predict a highly unsymmetrical transition state, and therefore a "two-stage" reaction[27a]. On the contrary, recent calculations by Leroy and coworkers favour the concerted mechanism[27b].

4.1. Concerted Pathway

The transition state (C) is found for $R_1 = R_2 = 2.21$ Å (Figure 6). At this point, the olefinic bond is optimized at 1.41 Å (still relatively close to a normal double bond) while the dienic moiety is entirely destroyed, the terminal bond (1.40 Å) being already longer than the

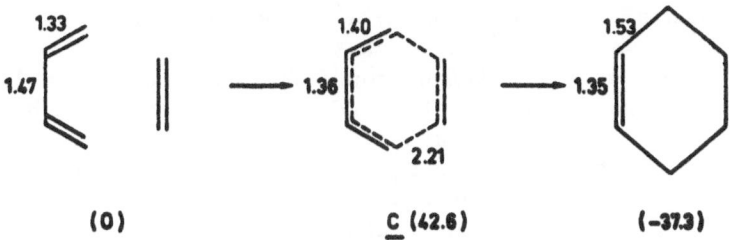

(0)   C (42.6)   (−37.3)

Figure 6 : Concerted mechanism for the Diels-Alder reaction. Bond lengths are in Angstrom.

central bond (1.36 Å). This indicates that the rearrangement of the butadiene skeleton occurs at a very early stage of the reaction but since the reaction is relatively exothermic ($\Delta H = -32$ kcal/mol), this result is somewhat in contradiction to Hammond's principle. The calculated activation energy, at the 4-31 G level, is 42.6 kcal/mol (exp. : 34.3 kcal/mol)[28].

4.2. Pure two-step pathway

Let us first consider the two conformations of the hex-2-ene 1,6 diyl diradical, labeled D_a and D_b, which should be primordial in such a pathway (figure 7). At the 4-31 G level, the energies of these

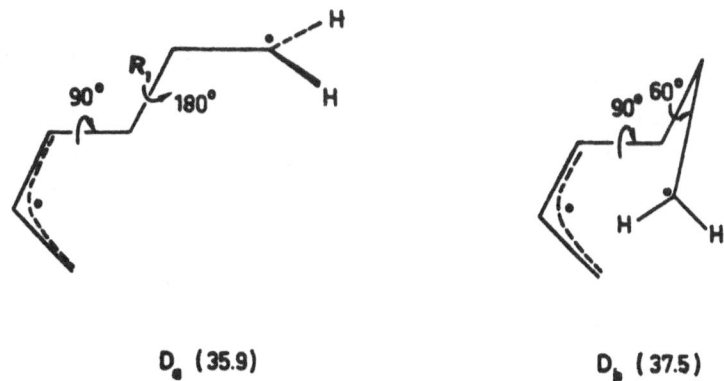

D_a (35.9) D_b (37.5)

Figure 7 : D_a and D_b conformations of the hex-2-ene 1,6 diyl diradical. The energies are in kcal/mol above that of reactants.

structures are 35.9 and 37.5 kcal/mol respectively. It is noteworthy that
these values are well below the calculated activation energy for the
concerted pathway (42.6 kcal/mol). Now consider the pure two-step pathway defined as first linking the reactants to give the relatively extended conformer D_a, D_a to D_b via internal rotation and, ultimately, a path
from D_b to cyclohexene.

$$\text{Reactants} \xrightarrow{R_1} D_a \to D_b \xrightarrow{R_2} \text{cyclohexene}$$

The small $D_a \to D_b$ step (rotation around the CC bond) has a low barrier
of 4.0 kcal/mol and the final step from D_b to cyclohexene has no activation energy. However, the formation of the first bond R_1 entails an
energy maximum which occurs for R_1 = 1.96 Å. The geometry of the transition state resembles that of D_a, except for R_1, and its energy lies
46.7 kcal/mol above that of the reactants. Thus the two main conclusions
are that i), the two-step pathway requires an activation energy 4.1 kcal/
mol higher than that of the concerted pathway and (ii), the energy
of the hex-2-ene 1,6 diyl diradical (D_a) lies 10.8 kcal/mol below the
two-step transition state. However, this diradical is not a secondary
minimum ; it appears that it must lie beyond the two step transition-
state,well on the side of the slope leading down to the cyclohexene.This
result clearly shows that the interpretation previously made of the
thermochemical calculations is based on the following implicit assumption : the diradical whose energy is calculated is the mid-point of the
reaction-path. With this assumption, the calculation of an enthalpy of
formation lower than the activation energy implies the existence of a
secondary minimum (Figure 8, on the left-hand-side).

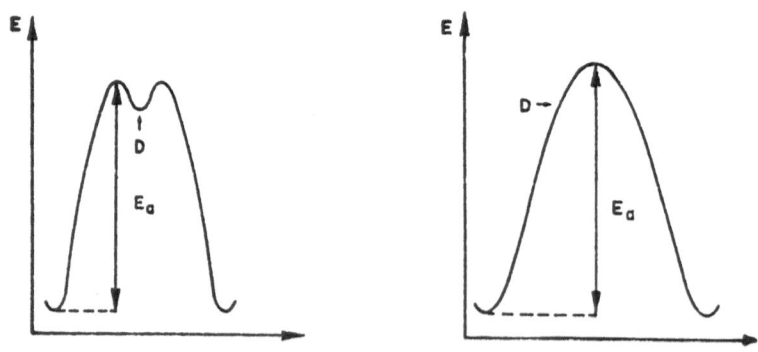

Figure 8 : Possible energy profiles when the estimated energy
of the diradical is below that of the transition state.

    However, if we are free from this hypothesis, it is no longer
true (Figure 8, on the right-hand-size). It is what we found is the

case of the Diels-Alder reaction, for which all the conformations of the diradical have an energy below that of the non-concerted transition state. This might also explain the apparent contradiction between the properties predicted by ab initio and thermochemical calculations for the trimethylene diradical. Indeed, we have shown that all the upper part of the potential energy curve corresponds to diradical structures, which differ only in their conformation. Consequently, there exists structures of the diradical, which are not secondary minima and whose energies are below that of transition state. These structures lie on both sides of the transition state.

4.3. Partially concerted pathway.

Finally, a partially concerted pathway has been examined, characterized by the following steps :

$$\text{reactants} \rightarrow C \rightarrow R \rightarrow D_b \rightarrow \text{cyclohexene}$$

This path begins with equal formation of both bonds ($\rightarrow$ C), and then continues by a rearrangement which shortens one bond further while lengthening the other bond anew ($\rightarrow D_b$). A monotonically descending path between C and D_b would imply that D_a can be reached via C with no further activation energy and would contradict the experimental, quasi-exclusive preference for a concerted pathway[26]. In fact, there is a new col R between C and D_b, which serves as a barrier preventing molecules in the concerted transition structure from freely rearranging to diradical. Its energy, 44.5 kcal/mol above reactants, is 1.9 kcal/mol above C. This difference is somewhat smaller than indicated by Bartlett's experiments[26] ($\sim$ 7.6 kcal/mol at 450° K), although the agreement would be excellent with the pure two-step pathway. A possibility is that two-step pathways through R involve a small additional dynamic barrier. Indeed, the sudden switch from a symmetric (R_1 and R_2 decreasing) stretching mode to an antisymmetric (R_1 decreasing, R_2 increasing) stretching mode implies a sharp right angle for the trajectories in multidimensional space and must be dynamically costly. The consequence of such a dynamic barrier might be to raise the effective activation energy of R at the level of the pure two-step pathway. The concerted pathway would then be favored by 4 kcal/mol, instead of $\sim$ 2 kcal/mol, in better agreement with the experiment.

5. THERMAL DECOMPOSITION OF 1-PYRAZOLINES

This extensively studied[29-33] decomposition reaction leads to cyclopropane and olefins and is related to the isomerization of cyclopropane since one of the most often invoked mechanisms involves the intermediacy of a trimethylene diradical.

In the case of pyrazolines labelled at positions 3 and 5 by alkyl groups[29-31], the major product is a cyclopropane with an apparent single inversion of stereochemistry. The first mechanism invoked to account for this inversion of stereochemistry, involved the simultaneous breaking

of the two carbon-nitrogen bonds, leading to a planar trimethylene
diradical 1[29b-e] which could easily cyclize by conrotatory motion
of both terminal groups[13] (scheme I). In order to test this mechanism,
Condit and Bergman studied the decomposition of bicyclic azo compounds

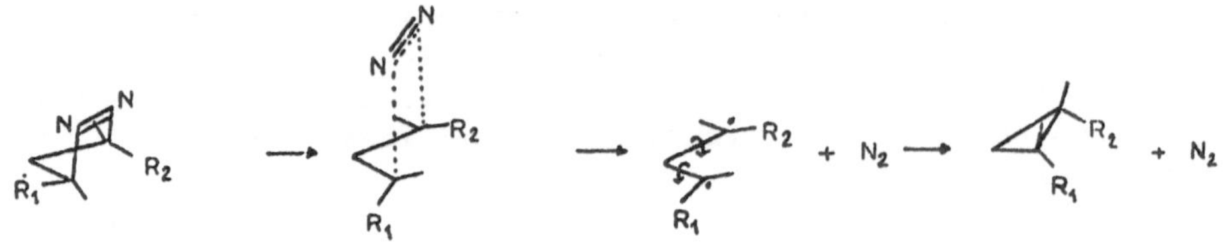

Scheme I

2-a and 2-b[31a]. In these molecules, a three-carbon bridge between carbons
3 and 4 should disfavour the formation of a planar diradical. However,
although it might be expected that simple extrusion of nitrogen, leading

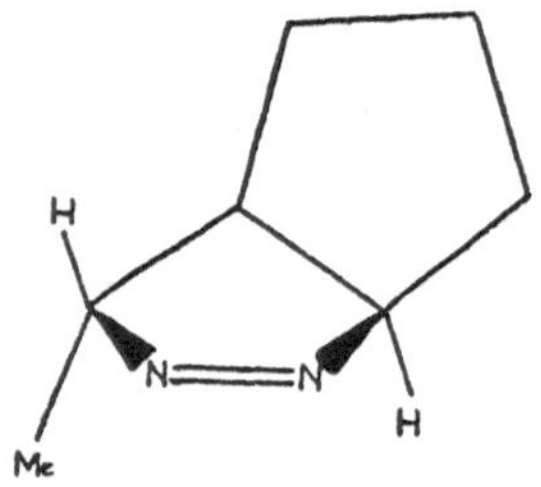

2-a 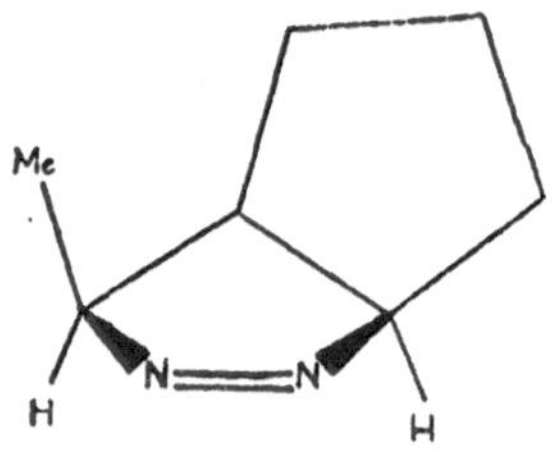   2-b

to retention of the stereochemistry, would now be the easiest pathway,
the major product still exhibits an inversion of stereochemistry. This
result led Condit and Bergann to propose an alternative mechanism, based
on an earlier suggestion by Roth and Martin[34]. In this second mechanism,
the slow step would be the breaking of only one carbon-nitrogen bond,
leading to the diradical 3. Then, the breaking of the second bond in
3 leads to a cyclopropane with an inverted stereochemistry (scheme III).
Later, other experimental data have supported this mechanism[35]. However,
in 1975, Clarke, Wendling and Bergman analysed the optical purity of
the products of the thermal decomposition of 3-ethyl-5-methyl-1-pyrazo-
lines[31e]. Their conclusion was that the product distribution cannot be

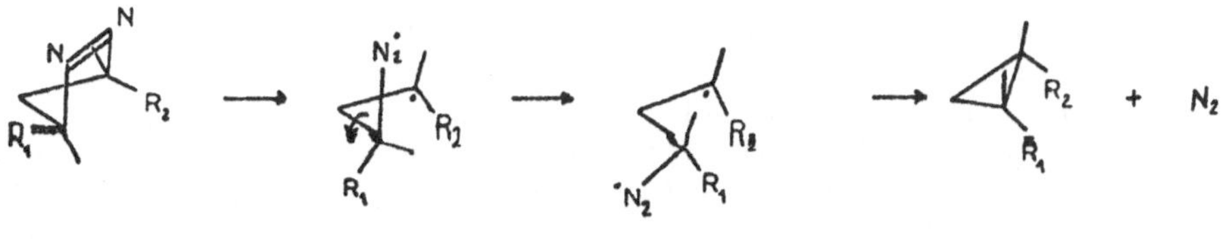

3

Scheme II

understood solely on the basis of the above mechanisms.

Our study is limited to the two limiting mechanisms already suggested to account for the inversion of stereochemistry, namely (a) synchronous or (b) successive breaking of the two carbon-nitrogen bonds.

5.1. Reaction paths involving synchronous breaking of both C-N bonds

For this mechanism, we have considered two limiting reaction paths:
the first one is described by Scheme I and leads to a cyclopropane with
the observed inversion of stereochemistry. The second path leads to
the "face-to-face" diradical 4 (Scheme III). Since there is no barrier
to reclosure from 4 to the cyclopropane[1], this reaction path gives

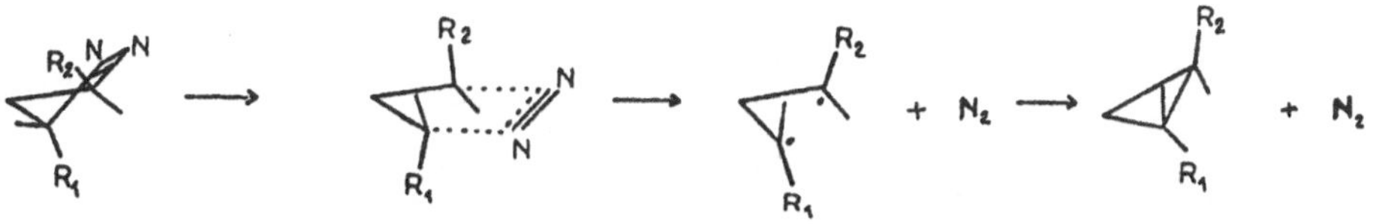

4

Scheme III

retention of stereochemistry. Considering the symmetry of the highest
(HO) and the lowest unoccupied (LU) molecular orbitals (MO) in the
diradicals 1 and 4, one can predict the first reaction path (Scheme I)
to be more favorable. Indeed, in the diradical 1, the HOMO is antisymmetrical with respect to the plane perpendicular to the carbon ring,
and the LUMO is symmetrical[13]. Thus, favorable interaction between the
HOMO (LUMO) of 1 and the π^* (π) orbital of nitrogen in the reverse
addition reaction can be expected. The inverse situation is observed
for the diradical 4, in which the HOMO is symmetrical while the LUMO
is antisymmetrical[13]. Thus, the formation of 1 is "allowed", while

that of 4 is "forbidden" in a concerted 2s+2s process. However, since
the frontier orbitals in diradicals are nearly degenerate, the energetic difference (after 3x3 CI) between the allowed and forbidden reaction paths must be rather small. Indeed, the calculated activation
energies, at the 4-31G level, for schemes I and III are 44.1 and 47.8
kcal/mol respectively (exp. 42.4 kcal/mol[29a]). The transition states
occur at C-N = 2.06 Å (Scheme I) and 2.07 Å (Scheme III).

5.2. The pure two-step pathway

    Consider the cleavage of the C_3-N_2 bond (Figure 9). Three main

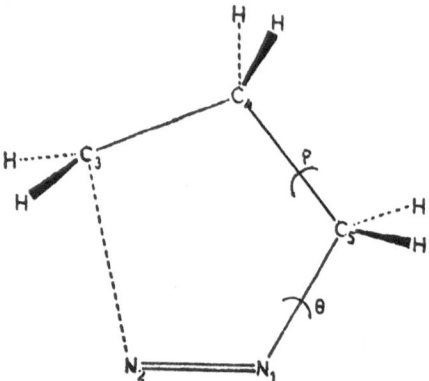

Figure 9 : Definition of the geometrical parameters used to
describe the breaking of the first C-N bond

geometrical parameters are used to describe this rupture : θ, rotational
angle of N_1-N_2 around the C_5-N_1 bond, ϕ, dihedral angle between the
$C_3C_4C_5$ and the $C_4C_5N_1$ planes, and the $<C_5N_1N_2$ angle. Note that diradical 3, involved in scheme II, corresponds to ϕ = 180° and to an un-
defined value of θ. The energetic profile, at the 4.31G level is pictured
in figure 10. The first step gives a transition state whose energy lies
44.5 kcal/mol above that of pyrazoline, occuring at a rotational angle
of ϕ = 120° (the value of θ is less critical, and the optimum value
is θ ≈ 120°). The most stable conformation of the nitrogen containing
diradical is obtained (figure 11) for ϕ=180°, and it is in agreement
with the backside displacement of N=N postulated in scheme II. The
second step (breaking of the second C-N bond) requires no further activation energy. Thus, the transition state for mechanism II is a structure
in which only one carbon-nitrogen bond is broken viz. a "pure" diradical,
and therefore, the nitrogen-containing diradical is not a secondary
minimum at the top of the non-concerted pathway.

    To summarize, we have shown that in the case of unsubstituted
pyrazolines, the two limiting reaction paths (i) simultaneous breaking
of both C-N bonds towards a coplanar diradical or (ii) successive

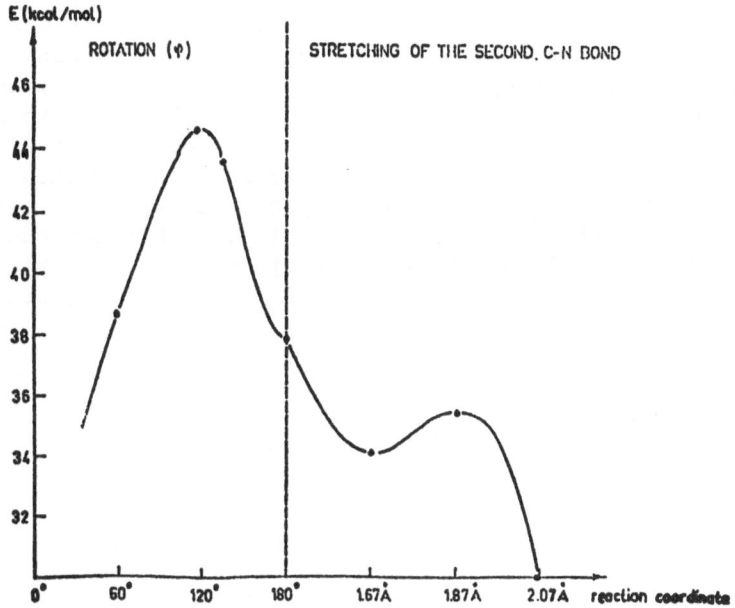

Figure 10 : Energy profile (at the 4-31 G level) for the
two-step mechanism. The energies are in kcal/mol above that
of pyrazoline.

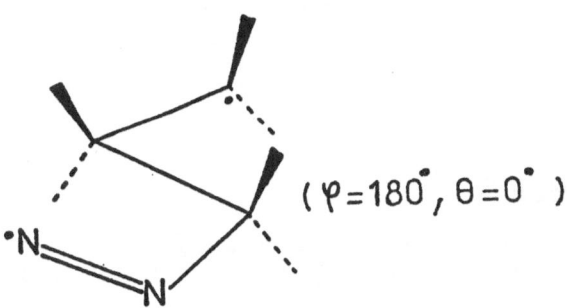

Figure 11 : Structure of the most stable conformation of the
nitrogen containing diradical.

breaking of the two C-N bonds via a nitrogen-containing diradical, require
approximately the same activation energy (44.1 and 44.5 kcal/mol respectively) and lead to a predominant single inversion of stereochemistry.

The activation energy of the third reaction path leading to retention
of the stereochemistry is 47.8 kcal/mol. However, it must be noted that
our calculations are not free of geometrical constraints. For instance,
in the two-step pathway, the rotation at the carbon radical center is
not allowed, and, in the first step, the NN bond length is fixed at
1.25 Å. The influence of these geometrical parameters will be tested
in further calculations[4]. Furthermore, an other mechanism, which
might well be competitive with the previous ones, has been suggested
by Nagaki and Fukui[35]. In this mechanism, both rotations and bond
cleavages are simultaneous, but not synchronous. Therefore, it is a
much more complex mechanism, which is under investigation.

CONCLUSION

To summarize, we have obtained for these thermal reactions, in the
same way as for others like the concerted elimination of hydrogen chloride from ethyl chloride[37] or ozonolysis reaction[38], a set of results
concerning the mechanisms which are qualitatively in agreement (at the
4-31G level) with experiment. Two points in particular - preferential
synchronous rotation of two groups in the trimethylene diradical, and
absence of a barrier for the reclosure of 1,3 diradicals - have received
experimental supports since the calculations were done. Furthermore,
these studies have led to a rather precise description of the properties
of the diradicals, which can be a transition state (reaction 1), a
secondary minimum (reaction 2) or a "transient point" on a hillside
(reaction 3). Finally, we believe that our results reconcile in a large
part the contradiction between the thermochemical estimates and the
quantum-mechanical calculations. They have shown that the diradicals
involved in the reaction schemes are not necessary the mid-point of
the pathway and can be located on one, or both sides of the transition
state.

ACKNOWLEDGEMENTS

The author thanks the organizing committee for the opportunity to
present this work at the 2e Séminaire International sur la Théorie
Quantique des Réactions Chimiques.

He also thanks Dr. R.C. Hayward for critical readings of the
manuscript.

The four reactions discussed in this paper have been studied by :
1) Y. Jean, L. Salem, J.S. Wright, J.A. Horsley, C. Moser and R.M.
Stevens ; 2) G.A. Segal ; 3) R.E. Townshend, G. Ramuni, G. Segal,
W.J. Hehre and L. Salem ; 4) P.C. Hiberty and Y. Jean.

REFERENCES

1. Y. Jean, L. Salem, J.S. Wright, J.A. Horsley, C. Moser and R.M. Stevens
Pure Appl. Chem., Suppl. 1, 197 (1971)
2. G.A. Segal, J. Amer. Chem. Soc. 96, 7892 (1974)
3. R.E. Townshend, G. Ramunni, G. Segal, W.J. Hehre and L. Salem, J.
Amer. Chem. Soc. 98, 2190 (1976)
4. P.C. Hiberty and Y. Jean, to be published
5. W.J. Hehre, W.A. Lathan, R. Ditchfield, M.D. Newton and J.A. Pople,
Program n° 236, Quantum Chemistry Program Exchange, Indiana University,
Bloomington, Ind.
6. A minimal basis set of Slater type orbitals has been used for this
reaction. R.M. Stevens, J. Chem. Phys. 52, 1397 (1970)
7. W.J. Hehre, R.F. Stewart and J.A. Pople, J. Chem. Phys. 51, 2657
(1969)
8. R. Ditchfield, W.J. Hehre and J.A. Pople, J. Chem. Phys. 54, 724
(1971)
9. R.K. Nesbet, Rev. Mod. Phys. 35, 552 (1963)
10. L. Salem and C. Rowland, Angew. Chem., Int. Ed. Engl. 11, 92 (1972)
11. B.S. Rabinovitch, E.W. Schlag and K.B. Wiberg, J. Chem. Phys. 28,
504 (1958)
12. F.T. Smith, J. Chem. Phys. 29, 235 (1958)
13. R. Hoffmann, J. Amer. Chem. Soc. 90, 1475 (1968)
14. S.W. Benson, J. Chem. Phys. 34, 521 (1961)
15. S.W. Benson, "Thermochemical Kinetics", John Wiley and Sons, New-York,
N.Y. (1968)
16. H.E. O'Neal and S.W. Benson, J. Phys. Chem. 72, 1866 (1968)
17. S.L. Buchwalter and G.L. Closs, J. Amer. Chem. Soc. 97, 3857 (1975)
18. J.A. Berson and L.D. Pedersen, J. Amer. Chem. Soc. 97, 238 (1975)
19. R.B. Woodward and R. Hoffmann, Angew. Chem.,Int. Ed. Engl. 8, 781
(1969)
20. J.S. Wright and L. Salem, J. Amer. Chem. Soc. 94, 322 (1972)
21. a) A.T. Coks, H.M. Frey and D.R. Stevens, Chem. Comm. 1969, 458
    b) J.E. Baldwin and P.W. Ford, J. Amer. Chem. Soc. 91, 7192 (1969)
22. F. Kern and W.D. Walters, J. Amer. Chem. Soc. 75, 6196 (1953)
23. S.W. Benson, J. Chem. Phys. 46, 4920 (1967)
24. In this study, a 15x15 configuration interaction follows the SCF
calculation. It includes all the single and double excitations from the
two highest occupied to the two lowest unoccupied molecular orbitals.
25. See reference 3 and references therein
26. P.D. Bartlett and K.E. Schueller, J. Amer. Chem. Soc. 90, 6071,
6077 (1968)
27. a) M.J.S. Dewar, A.C. Griffin and D. Kirschner, J. Amer. Chem. Soc.
96, 6225 (1974) ; b) L.A. Burke, G. Leroy and M. Sana, Theor. Chim. Acta
40, 313 (1975)

28. a) M. Uchiyama, T. Tomioha and A. Amano, J. Phys. Chem. 68, 1878
(1964) ; b) W. Tsang, J. Chem. Phys. 42, 1805 (1965) ; Int. J. Chem.
Kinet. 2, 311 (1970)
29. a) R.J. Crawford, R.J. Dummel and A. Mishra, J. Amer. Chem. Soc. 87,
3023 (1965) ; b) R.J. Crawford and A. Mishra, J. Amer. Chem. Soc. 87, 3768
(1965) ; c) R.J. Crawford and A. Mishra, J. Amer. Chem. Soc. 88, 3959
(1966) ; d) R.J. Crawford and G.L. Erickson, J. Amer. Chem. Soc. 89,
3907 (1967) ; (e) R.J. Crawford and L.H. Ali, J. Amer. Chem. Soc. 89,
3908 (1967) ; (f) A. Mishra and R.J. Crawford, Can. J. Chem. 47, 1515
(1969) ; g) M.P. Schneider and R.J. Crawford, Can. J. Chem. 48, 628
(1970) ; h) R.J. Crawford and M. Ohno, Can. J. Chem. 52, 3134 (1974) ;
i) R.J. Crawford and H. Tokunaga, Can. J. Chem. 52, 4033 (1974)
30. a) D.E. McGreer, N.W.K. Chiu, M.G. Vinje and K.C.K. Wong, Can. J.
Chem. 43, 1407 (1965) ; b) D.E. McGreer and W.S. Wu, Can. J. Chem. 45,
461 (1967)
31. a) P.B. Condit and R.G. Bergman, Chem. Comm. 4 (1971); b) D.H. White,
P.B. Condit and R.G. Bergman, J. Amer. Chem. Soc. 94, 7931 (1972) ;
c) R.A. Keppel and R.G. Bergman, J. Amer. Chem. Soc. 94, 1350 (1972) ;
d) R.G. Bergamn in "Free Radicals", Vol. 1, J. Kochi, Ed. Wiley, New
York, N.Y., 1973, Chapter 5 ; e) T.C. Clarke, L.A. Wendling and R.G.
Bergman, J. Amer. Chem. Soc. 97, 5638 (1975)
32. C.G. Overberger and J.P. Anselme, J. Amer. Chem. Soc. 84, 869 (1962);
b) C.G. Overberger, J.P. Anselme and J.R. Hall, J. Amer. Chem. Soc. 85,
2752 (1963) ; c) C.G. Overberger and J.P. Anselme, J. Amer. Chem. Soc.
86, 658 (1964) ; d) C.G. Overberger, N. Weinshenker and J.P. Anselme,
J. Amer. Chem. Soc 86, 5364 (1964) ; e) C.G. Overberger, N. Weinshenker
and J.P. Anselme, J. Amer. Chem. Soc. 87, 4119 (1965)
33. M. Schneider and H. Strohäcker, Tetrah. 32, 619 (1976)
34. a) W.R. Roth and M. Martin, Tetrah. Let. 4695 (1967) ; b) W.R. Roth
and M. Martin, Annalen 702, 1 (1967)
35. T. Sasaki, S. Eguchi and F. Hibi, Chem. Comm. 227 (1974)
36. S. Inagaki and K. Fukui, Bull. Chem. Soc. Japan 45, 824 (1972)
37. P.C. Hiberty, J. Amer. Chem. Soc. 97, 5975 (1975)
38. P.C. Hiberty, J. Amer. Chem. Soc. 98, 6088 (1976)

THEORETICAL STUDIES OF SN_2 REACTIONS.

A. Dedieu and A. Veillard
E.R. n° 139 du CNRS
Université Louis Pasteur
Strasbourg, France

INTRODUCTION

The nucleophilic substitution is an heterolytic reaction of the type

$$Y: + R : X \longrightarrow Y : R + :X \qquad (1)$$

Y is called the nucleophile (usually an anion or a neutral base), X is called the leaving group. The atom which undergoes the substitution (usually a carbon atom in what follows) is called the electrophilic center. Two mechanisms are generally considered for the nucleophilic substitutions [1] :
- the SN_1 unimolecular substitution, which takes place in two steps :

$$CR_3Br \rightleftharpoons CR_3^+ + Br^- \qquad (2)$$
$$CR_3^+ + OH^- \longrightarrow CR_3OH \qquad (3)$$

the first step corresponding to a slow and reversible dissociation to a cationic species. This mechanism usually leads to the racemization of an optically active substrate.
- the SN_2 bimolecular substitution, which is a single step reaction. In the mechanism postulated by Ingold [2], the backside approach of the nucleophile and the loss of the leaving group proceed in a concerted way. A single transition state is found along the minimum energy path with the electrophilic center bound to both the nucleophile and the leaving group (Fig. 1). A second order reaction rate ($v = k$ [Y] [RX]) found for this type of reaction supports the proposed mechanism. A SN_2 substitution proceeds with inversion of configuration (the reaction is sometimes called the Walden inversion). Optical isomerization has been reported experimentally [3] and also supports the proposed mechanism.

For a given substitution reaction, the assignment of the mechanism as SN_1 or SN_2 is usually based on the kinetics of the reaction. However,

R. Daudel, A. Pullman, L. Salem, and A. Veillard (eds.), Quantum Theory of Chemical Reactions, Volume I, 69–89.

it is sometimes difficult to choose between the two mechanisms SN_1 or SN_2 since there are borderline cases, with the possibility that both me-chanisms operate simultaneously [4]. Various attempts have been made to unify the proposed mechanisms for the nucleophilic substitutions [4-7]. Most recent, the Sneen ion-pair mechanism [8] is based on the theory of ion pairs proposed by Winstein [9-11]. According to this author, the follo-wing equilibrium takes place in solution

$$R - X \rightleftharpoons R^+X^- \rightleftharpoons R^+ \quad X^- \rightleftharpoons R^+ + X^- \qquad (4)$$
$$\quad I \qquad\quad II \qquad\quad III \qquad\quad IV$$

where II and III represent ion-pairs (intimate ion pair and solvent-sepa-rated ion pair). The nucleophile Y may react with either of the four spe-cies I-IV according to the generalized mechanism

$$RX \underset{k_{-1}}{\overset{k_1}{\rightleftharpoons}} R^+X^- \underset{k_{-2}}{\overset{k_2}{\rightleftharpoons}} R^+ \quad X^- \underset{k_{-3}}{\overset{k_3}{\rightleftharpoons}} R^+ + X^- \qquad (5)$$
$$\downarrow k_4 \qquad \downarrow k_5 \qquad\quad \downarrow k_6 \qquad\qquad \downarrow k_7$$
$$RY \qquad RY \qquad\quad RY \qquad\qquad RY$$

According to Sneen [12] and others [13], the step k_4, namely the attack of RX by Y, does not occur usually. The SN_2 reaction would rather result from the steps k_1 and k_5. A SN_1 mechanism corresponds to the steps k_1, k_2, k_3 and k_7.

A goal for many experimental studies of SN_2 reactions was the esta-blishement of a nucleophilicity scale, such as the one of Edwards and Pearson [14] (a good nucleophile being characterized by a fast rate of reac-tion). Edwards and Pearson proposed the following scale of nucleophilici-ty

$$I^- > CN^- > SCN^- > OH^- > N_3^- > Br^- > C_6H_5O^- > Cl^- > H_2O$$

where the nucleophilicity of the reagent depends on both its basicity and its polarizability. However it turned out frequently that the scale of nucleophilicity for a series of reagents is very sensitive to solvent changes. The following order of nucleophilicity with methyl alcohol as solvent (protic solvent) [15]

$$I^- > SCN^- \simeq CN^- > Br^- > Cl^-$$

becomes in an aprotic solvent [16]

$$CN^- > Cl^- \simeq Br^- > I^- > SCN^-$$

On the other hand, the reactivity of a given substrate RX is deter-mined largely by the nature of the leaving group X. The ease of expulsion of the leaving group has been referred as the leaving group ability. Strong bases behave usually as poor leaving groups [17]. According to Pearson [18], the ease of expulsion of the leaving group depends also on "symbiotic effects", namely the substitution is easier when both the nucleophile and the leaving group are either "hard bases" or "soft bases" according to the classification of Klopman [19].

The changes introduced when passing from a polar protic solvent to a dipolar aprotic solvent has been a source of information on the structure of the transition state. Measurements of the enthalpies of transfer (from methanol to dimethylformamide) of the reactants ($\delta \Delta H_s^r$), of the transition states δH^t and of the products ($\delta \Delta H_s^p$) for a series of SN₂ reactions were reported by Haberfield [20]. The ratio $\delta H^t/(\delta \Delta H_s^p - \delta \Delta H_s^r)$ was used as a measure of the resemblance of the transition state to the reactants and to the products, with the following conclusions : i) increasing the basicity of the nucleophile moves the transition state structure closer to that of the reactants ; ii) increasing the basicity of the leaving group moves the transition state closer to the products ; iii) electron withdrawal on the carbon undergoing substitution moves the transition state closer to the products ; iv) other factors being equal, in SN₂ reactions having a neutral nucleophile and a neutral electrophile the transition state resembles the reactants more than is the case in SN₂ reactions having a negative nucleophile and a neutral electrophile.

Until 1971, the experimental work on the SN₂ reaction was limited to studies made in solution. Since that time, a number of experimental studies of SN₂ reactions in the gas phase have been reported [21-25]. Bohme et al have studied the reactions [22]

$$X^- + CH_3CL \longrightarrow CH_3X + CL^- \qquad X = H, NH_2, OH, F, C_2H, CN \qquad (6)$$

$$X^- + CH_3F \longrightarrow CH_3X + F^- \qquad X = H, NH_2, OH, C_2H, CN \qquad (7)$$

CL^- and F^- ions were the only products observed in all of these reactions which should correspond to nucleophilic displacement. It turned out that the nucleophilicity scale depends on the substrate, being

$$H^- > NH_2^- > OH^- \simeq F^- > C_2H^- > CN^-$$

for the reaction with CH_3CL and

$$OH^- > NH_2^- \simeq H^- > C_2H^- \simeq CN^-$$

for the reaction with CH_3F. The observed nucleophilicity does not parallel the basicity in the gas phase (the scale of basicity being [26] $NH_2^- > H^- > OH^- > F^-$). Lieder and Brauman have confirmed that, for a series of reactions

$$X^- + CH_3Y \longrightarrow Y^- + CH_3X \qquad (8)$$

the neutral product is indeed CH_3X [23]. For the same reaction with
$X = F, CL, CH_3S$ and $Y = CL, BR$ [24], the nucleophilic order $F^- > CH_3S^- > CL^-$
was maintained for both methyl chloride and bromide, however the relative
nucleophilicity $F^- : CH_3S^- : CL^-$ changes from $10 : 1 : 0.1$ toward methyl
chloride, to $8 : 2 : 1$ toward methyl bromide. Even more striking was the
reversal in leaving group ability (LGA) dependence on the nucleophile,
the relative LGA of $CL^- : BR^-$ being $10 : 8$ with F^- as the nucleophile,
$1 : 2$ with CH_3S^- and $1 : 10$ with CL^- [24]. The reaction

$$CL^- + CH_3CL \longrightarrow CH_3CL + CL^- \qquad (9)$$

turned out to be much slower than the other reactions of the type (8)
studied, thus ruling out a symbiotic effect [24]. The same authors found
that the displacement reaction of CL^- on the cis- and trans-4-bromocyclo-
hexanols occurs with configuration inversion [23], thus supporting a reaction mechanism which corresponds to Walden inversion and backside attack.

A stimulus to the gas-phase studies of SN_2 reactions was the availability of a number of theoretical calculations of the corresponding potential energy surfaces (cf. below the list of such calculations). These calculations were usually carried out for the reactions proceeding in vacuo,
i.e. in the absence of solvent. There were many reasons for carrying out
theoretical studies of SN_2 reactions, but the main one was to try to iso-
late intrinsic properties of the reactants and the products, free of solvent effects. Among the other reasons, some being rather theoretical in
nature while others were more closely related to experimental aspects,
one may quote :
    i) the unexpected result of the first theoretical study [27], with
the "transition state" $\left[F \cdots CH_3 \cdots F \right]^-$ for the reaction

$$F^- + CH_3F \longrightarrow CH_3F + F^- \qquad (10)$$

found more stable than the reactants and the products (Fig. 2). For the
similar reactions

$$CL^- + CH_3CL \longrightarrow CH_3CL + CL^- \qquad (11)$$

$$BR^- + CH_3BR \longrightarrow CH_3BR + BR^- \qquad (12)$$

experimental activation energies of 20 and 16 kcal/mole respectively have
been reported with acetone as solvent [28]. The difference, of at least
20 kcal/mole, between the negative energy barrier computed for (10) and
the estimated activation energy (based on the values reported for (11)
and (12)) was assigned to the change in solvation energy between the reactants and the transition state (this was not unrealistic since a value
of 3-4 kcal/mole for the activation energy of reaction (11) in the gas
phase has been estimated [29] from the data of Ref. [24] and may be compared
with the above value of 20 kcal/mole in solution). However, these theoretical results raised the possibility of stable five-coordinate carbon
compounds. Systems such as CBR_5^- had been reported [30] but their structure

was unknown at that time [31].

    ii) ab initio calculations of potential energy surfaces for reactions of chemical interest are usually carried out, at best, at the SCF level, while the correlation effects are neglected. It is commonly assumed for some classes of reactions (whenever the number of closed and open shells remains constant throughout the reaction) that the correlation energy may be considered as approximately constant along the reaction path. SN$_2$ reactions are obvious candidates for testing this proposal.

    iii) according to a proposal by Gillespie and Ugi based on CNDO calculations, the symmetrical configuration $[FCH_3F]^-$ on the reaction path would no longer be a transition state but rather an intermediate (Fig.3)[32]. This intermediate would have a sufficient lifetime to allow intramolecular rearrangements to occur (either by the Berry pseudorotation [33] or by the turnstile mechanism [34]). In this way the reaction might lead to retention of configuration.

    In what follows, our discussion of the theoretical studies of SN$_2$ reactions will be restricted to the reactions occuring at a saturated carbon center.

THE CALCULATIONS

    Table 1 summarizes the ab initio calculations of potential energy surfaces for SN$_2$ reactions [27,35-45]. For many of the reactions studied, such as

$$H^- + CH_4 \longrightarrow CH_4 + H^- \qquad\qquad (13)$$

$$F^- + CH_3F \longrightarrow CH_3F + F^- \qquad\qquad (14)$$

the products and the reactants are identical. It turns out that, for these reactions, the transition state is symmetrical with respect to the plane passing through the carbon atom and the three hydrogen atoms (namely the mid-point of the reaction represents the transition state and not an intermediate as in Fig. 3). Then finding the geometry of the transition state requires only the optimization of two parameters, namely the C-H bond length and the C$\cdots$X distance. A reaction such as

$$H^- + CH_3F \longrightarrow CH_3F + H^- \qquad\qquad (15)$$

was included in order to study the influence of an electron withdrawing substituent on the electrophile center.

    Since the discussion will be devoted to the chemical bearings of these calculations, two features of the calculations should be underlined :

    i) we have mentioned above that, in the first calculation for the reaction (14), no energy barrier was found, namely the mid-point of the [27] reaction turned out to be more stable than the reactants or the products[27]. A number of calculations have reached the same conclusion [35,37,42,45]. However this is an artefact of the calculation since it is known, both theoretically from more refined calculations [35,37,43] and experimentally [22,24], that this type of reactions requires an activation energy. This artefact is common to all the calculations which lack simultaneously the d-type polarization functions on the electrophilic carbon and the p-type diffuse functions on the incoming and leaving fluorine atoms [35] (these diffuse functions insure a balanced calculation of the reactants including the anion F$^-$ and of the transition state where the negative charge is delocalized on several atoms) (Table II). Nevertheless when a flexible s,p basis set is used the mid-point turns out to be a local transition state, i.e. higher in energy than some intermediate points on the reaction path but lower in energy than the reactants [27,37,45].

ii) the discussion will be based on the results of the calculations at the SCF level (comparable results for a series of reactions are available only at the SCF level). Since the discussion will be largely focused on the change in energy along the reaction path, we shall assume that the correlation energy remains approximately constant along this path. We have reported in Table III the change in correlation energy along the reaction path given by a number of configuration interaction (CI) calculations. For the SN_2 reactions, this change does not exceed probably a few kcal/mole (the large change in correlation energy of 15 kcal/mole reported by Dyczmons and Kutzelnigg for reaction (13) [48] is probably an artefact of the method [43]).

Besides the ab initio calculations, one should also mention a number of semi-empirical calculations [32,46,49-54], mostly with the CNDO method. However some care should be exercised when using the results of the CNDO calculations since :

i) the optimized bond lengths from the CNDO calculations differ appreciably from the corresponding ones given by the ab initio calculations. For the transition state CH_5^- the CNDO calculation yields for the axial bond length a value of 1.21 Å [52] versus values in the range 1.63 - 1.74 Å from the ab initio calculations [36,38,55,56]. The optimized value of the C-F bond length for the transition state $[FCH_3F]^-$ is 1.44 Å from the CNDO calculation [51] versus values in the range 1.82 - 1.88 Å from the ab initio calculations [27,37,38].

ii) the relative energies along the reaction path appear to be seriously in error. For reaction (14), the mid-point of the reaction is stabilized with respect to the reactants by 88 kcal/mole [51] whereas it appears destabilized in the most accurate ab initio calculations (see Table II). A similar stabilization of the symmetric "transition states" is also found in Ref. [54].

iii) the CNDO calculations fail to predict the relative stabilities of the five-coordinate systems : the C_{2v} configuration of $[FCH_3F]^-$ with two fluorine atoms as equatorial ligands is found more stable than the D$_{3h}$ configuration with the fluorine ligands axial [32] whereas, in the ab initio calculations [45], the C$_{2v}$ structure is not stable and rearranges directly to the D$_{3h}$ structure.

On the contrary, the Extended Hückel calculations [49] perform relatively well with respect to the optimized bond lengths and the relative energies when compared to the results of the ab initio calculations.

DISCUSSION

Energy profiles along the reaction path corresponding to the usual backside attack have been reported for reactions (13)(14)(15) and (16) [38,39,45]

$$H^- + CH_3F \longrightarrow CH_4 + F^- \qquad (16)$$

(Fig. 4). When the nucleophile approaches the substrate, the energy decreases first below its value for the reactants and goes through a minimum before raising to a maximum which corresponds to the transition state. The minimum corresponds to a weak complex of C$_{3v}$ symmetry (Fig. 5) stabilized by a charge-dipole interaction, at a distance between the nucleophile and the electrophilic center in the range 2.5 - 3.5 Å. Such cluster ions formed by the interaction of an hydrocarbon and an halide ion have been reported in the gas phase by ion cyclotron resonance and mass spectroscopy [57-60]. The binding energy of 13.2 kcal/mole calculated for the cluster (FCH$_3$)F$^-$ [39] may be compared to the experimental enthalpy of association of 8.6 kcal/mole for the cluster (ClCH$_3$)Cl$^-$ [58]. Another ion-molecule cluster (FCH$_3$)F$^-$ with the fluorine ion bound to a single hydrogen may be as stable as the cluster of C$_{3v}$ symmetry.

The other reaction paths, corresponding for instance to a frontside attack, have been considered by Schlegel et al for reaction (14) [45]. Path 1 of Fig. 6 corresponds to the backside attack already considered. Path 2 along the bisector of H$_4$CH$_5$ should lead to a C$_{2v}$ trigonal bipyramidal transition state I with the two fluorine atoms equatorial. However this structure is not stable and optimizes directly to the D$_{3h}$ transition state II (taken as the energy reference in Fig. 6). Path 3, corresponding to the attack on the 2H,F face, induces a SN$_2$ displacement of an hydride ion by a fluoride ion via normal backside attack

$$F^- + CH_3F \longrightarrow H^- + CH_2F_2 \qquad (17)$$

and proceeds through the transition state III which is energetically unfavorable. Path 4 following the bisector of H$_3$CF$_2$ goes through a C$_s$ transition state IV and leads to the SN$_2$ displacement of fluoride with retention of configuration but is again energetically unfavorable. Path 5 with the fluoride ion approaching a direction perpendicular to a HCF plane proceeds through a transition state V which is the true transition state for the displacement with retention. The frontside attack appears energetically less favorable than the backside attack by about 56 kcal/mole.

We have reported in Table IV the calculated values of the energy barrier for a number of SN_2 reactions together with the corresponding activation energies for the gas-phase reactions. The agreement appears rather satisfactory considering i) the rather crude approximations needed in order to get an estimate of the activation energy ; ii) the fact that the theoretical barriers may be in error by a few kcal/mole ; iii) the distinction between activation energies obtained from the Arrhenius equation and energy barriers.

The calculated energy barriers provide a basis for testing a number of proposals which relate the ease of the reaction to the nature of the reactants and the products :

i) comparable energy barriers of respectively 3.8 and 7.9 kcal/mole have been reported for the two reactions

$$H^- + CH_3F \longrightarrow CH_4 + F^- \qquad\qquad (18)$$

$$F^- + CH_3F \longrightarrow CH_3F + F^- \qquad\qquad (19)$$

although for the first reaction the nucleophile is a soft base and the leaving group a hard base, whereas for the second one both the nucleophile and the leaving group are hard bases. These results appear to rule out a symbiotic effect.

ii) in Table V reactions have been classified according to the nature of the leaving group, H^-, F^- or Cl^-. The highest barriers correspond to the reactions with H^- as leaving group and the lowest ones to the reactions with Cl^- as leaving group. Thus increasing the basicity of the leaving group appears to make its displacement more difficult (we rely on the gas-phase basicity order $H^- > F^- > Cl^-$ with the respective proton affinities of 400, 370 and 335 kcal/mole [29]). However this correlation is probably a limited one since, in the following reactions

$$F^- + CH_3F \longrightarrow CH_3F + F^- \qquad\qquad (20)$$

$$F^- + CH_3CN \longrightarrow CH_3F + CN^- \qquad\qquad (21)$$

F^- is displaced more easily than CN^- (the computed energy barriers for (20) and (21) being respectively 7 and 17 kcal/mole [37]) although F^- has a greater proton affinity than CN^- (respectively 370 and 340 kcal/mole).

iii) each series of reactions in Table V correspond to different nucleophiles acting on a given substrate (CH_4, CH_3F or CH_3Cl) with a given leaving group (H^-, F^- or Cl^-). A common order of nucleophilicity $H^- > F^- > CN^- > Cl^-$ is found for the three series, which corresponds to the sequence of proton affinities. One will notice that CN^- is a poorer nucleophile than F$^-$ despite a much greater polarizability.

Theoretical calculations of potential energy surfaces not only yield the energy barrier and the exothermicity of a reaction, buth they also provide us with the structure of the transition state which is not accessible otherwise, making possible to test the proposed structural relationships between the transition state, the reactants and the products :

i) in agreement with the Hammond postulate, the transition state (FCH$_4$)$^-$ for the exothermic reaction

$$H^- + CH_3F \longrightarrow CH_4 + F^- \tag{22}$$

appears closer to the reactant CH$_3$F than to the product CH$_4$, since the increase for the C-F bond length from the reactant CH$_3$F to the transition state (FCH$_4$)$^-$ is 0.54 Å whereas the decrease in the C-H bond length from the transition state to the product CH$_4$ is 0.86 Å [39] (Fig. 7).

ii) for the transition states (CH$_4$X)$^-$ (X = H,F,Cl) corresponding to the reaction

$$X^- + CH_4 \longrightarrow CH_3X + H^- \tag{23}$$

the bond length C-H (corresponding to the leaving group) increases within the sequence H < F < Cl (Table VI), in other words the leaving group moves further away when the basicity of the nucleophile decreases. This is in agreement with the proposal by Haberfield that an increase in the basicity of the leaving group moves the transition state closer to the product [20]. However the same correlation breaks down (Table VI) for the C-F bond length in the transition states (XCH$_3$F)$^-$ corresponding to the reaction

$$X^- + CH_3F \longrightarrow CH_3X + F^- \tag{24}$$

(possibly as a consequence of a lack of accuracy in the determination of the C-F bond length for the transition state (FCH$_4$)$^-$ [39]).

iii) the axial C-H bond length decreases from 1.74 Å in the transition state (CH$_5$)$^-$ of reaction (13) to 1.60 Å in the transition state (CFH$_4$)$^-$ for reaction (15). Thus the replacement of a nonreacting hydrogen atom by an electronwithdrawing substituent brings about a shortening of the reacting bonds in the transition state. This is in agreement with a prediction by Thornton that an electron-withdrawing substituent must cause a shortening of the reacting bonds in a symmetrical transition state [62,63]

CONCLUSION

Ab initio quantum mechanical calculations provide probably a satisfactory description of SN$_2$ reactions _in vacuo_, with respect to both the

energetics of the reactions and the structure of the transition states.
The agreement with the data obtained from   gas-phase experiments appears
satisfactory. The results are coherent with a concerted mechanism and
the occurence usually of a transition state. Future studies should investigate to which extent the description reached for the reactions in vacuo
can be extended to the reactions in solution. A preliminary attempt along
this line has been made by Minot and Trong Anh, who have related the
change in nucleophilicity of the halide ions when going from an aprotic
solvent to a protic one to the change in orbital energy of the highest
occupied orbital [64]. Theoretical studies of the exact mechanism of SN_2
reactions in solution should probably start with a study of the concept
of ion-pair.

REFERENCES

[1]   E.D. Hughes and C.K. Ingold, J. Chem. Soc., 236 (1935).
[2]   C.K. Ingold, "Structure and Mechanism in Organic Chemistry", New-
      York, Cornell University Press, 1953.
[3]   H. Phillips, J. Chem. Soc., 123, 44 (1923) ; 127, 2552 (1925).
[4]   V. Gold, J. Chem. Soc., 4633 (1956).
[5]   C.G. Swain, J. Amer. Chem. Soc., 70, 1119 (1948).
[6]   C.G. Swain and R.W. Eddy, J. Amer. Chem. Soc., 70, 2989 (1948).
[7]   S. Winstein, E. Grunwald and H.W. Jones, J. Amer. Chem. Soc., 73,
      2700 (1951).
[8]   R.A. Sneen, Acc. Chem. Res., 6, 46 (1973).
[9]   S. Winstein, P.E. Klinedist and G.C. Robinson, J. Amer. Chem. Soc.,
      83, 885 (1961).
[10] S. Winstein, P.E. Klinedist and E. Clippinger, J. Amer. Chem. Soc.,
      83, 4986 (1961).
[11] S. Winstein, R. Baker and S.G. Smith, J. Amer. Chem. Soc., 86, 2072
      (1964).
[12] R.A. Sneen and J.W. Larsen, J. Amer. Chem. Soc., 91, 6031 (1969).
[13] J.M.W. Scott, Can. J. Chem., 48, 3807 (1970).
[14] J.O. Edwards and R.G. Pearson, J. Amer. Chem. Soc., 84, 16 (1962).
[15] R.G. Pearson, H. Sobel and J. Songstad, J. Amer. Chem. Soc., 90,
      319 (1968).
[16] A.J. Parker, Chem. Rev., 69, 1 (1969).
[17] C.A. Bunton, "Reaction mechanisms in organic chemistry", J.H. Ridd
      ed., Vol. 1, Elsevier, Amsterdam, 1963.
[18] R.G. Pearson and J. Songstad, J. Org. Chem., 32, 2899 (1967).
[19] G. Klopman, J. Amer. Chem. Soc., 90, 223 (1968).
[20] P. Haberfield, J. Amer. Chem. Soc., 93, 2091 (1971).
[21] L.R. Young, E. Lee-Ruff and D.K. Bohme, Chem. Commun., 35 (1973).
[22] D.K. Bohme, G.I. Mackay and J.D. Payzant, J. Amer. Chem. Soc., 96,
      4027 (1974).
[23] C.A. Lieder and J.I. Brauman, J. Amer. Chem. Soc., 96, 4028 (1974).
[24] J.I. Brauman, W.N. Olmstead and C.A. Lieder, J. Amer. Chem. Soc.,
      96, 4030 (1974).

[25] J.D. Payzant, K. Tanaka, L.D. Betowski and D.K. Bohme, J. Amer. Chem. Soc., 98, 894 (1976).

[26] L.B. Young, E. Lee-Ruff and D.K. Bohme, Canad. J. Chem., 49, 979 (1971).

[27] G. Berthier, D.-J. David and A. Veillard, Theoret. Chim. Acta, 14, 329 (1969).

[28] P.B.D. de la Mare, J. Chem. Soc., 3169, 3180 (1955).

[29] A. Dedieu, Thèse de Doctorat d'Etat, Strasbourg, 1975.

[30] F. Effenberger, W.D. Stohrer and A. Steinbach, Angew. Chem., 81, 261 (1969).

[31] H.F. Lindner and B. Kitsche-von Gross, Chem. Ber., 109, 314 (1976).

[32] P.D. Gillespie and I. Ugi, Angew. Chem. Int. Ed., 10, 503 (1971).

[33] R.S. Berry, J. Chem. Phys., 32, 993 (1960).

[34] I. Ugi, D. Marquading, H. Klusacek and P. Gillespie, Acc. Chem. Res., 4, 288 (1971).

[35] A. Dedieu and A. Veillard, Chem. Phys. Let., 5, 328 (1970).

[36] C.D. Ritchie and G.A. Chappell, J. Amer. Chem. Soc., 92, 1819 (1970).

[37] A.J. Duke and R.F.W. Bader, Chem. Phys. Let., 10, 631 (1971).

[38] A. Dedieu and A. Veillard, in "Reaction transition states", J.-E. Dubois ed., Gordon and Breach Science Publishers, New-York, 1972.

[39] A. Dedieu and A. Veillard, J. Amer. Chem. Soc., 94, 6730 (1972).

[40] R.F.W. Bader, A.J. Duke and R.R. Messer, J. Amer. Chem. Soc., 95, 7715 (1973).

[41] A. Dedieu, A. Veillard and B. Roos, in "Chemical and biochemical reactivity", E.D. Bergmann and B. Pullman ed., The Israel Academy of Sciences and Humanities Jerusalem, 1974, p. 371.

[42] P. Baybutt, Mol. Phys., 29, 389 (1975).

[43] F. Keil and R. Ahlrichs, J. Amer. Chem. Soc., 98, 4787 (1976).

[44] P. Cremaschi and M. Simonetta, Chem. Phys. Let., 44, 70 (1976).

[45] H.B. Schlegel, K. Mislow, F. Bernardi and A. Bottoni, Theoret. Chim. Acta, 44, 245 (1977).

[46] W.-D. Stohrer and K.R. Schmieder, Chem. Ber., 109, 285 (1976).

[47] A. Macias, J. Chem. Phys., 49, 2198 (1969).

[48] V. Dyczmons and W. Kutzelnigg, Theoret. Chim. Acta, 33, 239 (1974).

[49] D.R. Kelsey and R.G. Bergman, J. Amer. Chem. Soc., 93, 1953 (1971).

[50] J.P. Lowe, J. Amer. Chem. Soc., 93, 301 (1971).

[51] P. Cremaschi, A. Gamba and M. Simonetta, Theoret. Chim. Acta, 25, 237 (1972).

[52] N.L. Allinger, J.C. Tai and F.T. Wu, J. Amer. Chem. Soc., 92, 579 (1970).

[53] W.-D. Stohrer, Chem. Ber., 107, 1795 (1974).

[54] J.J. Dannenberg, J. Amer. Chem. Soc., 98, 6261 (1976).

[55] W. Th. Van der Lugt and P. Ros, Chem. Phys. Let., 4, 389 (1969).

[56] J.J.C. Mulder and J.S. Wright, Chem. Phys. Let., 5, 445 (1969).

[57] J.M. Riveros, A.C. Breda and L.K. Blair, J. Amer. Chem. Soc., 95, 4066 (1973).

[58] R.C. Dougherty, J. Dalton and J.D. Roberts, Org. Mass Spectr., 8, 77 (1974).

[59] R.C. Dougherty and J.D. Roberts, Org. Mass Spectr., 8, 81 (1974).

[60] R.C. Dougherty, Org. Mass Spectr., 8, 85 (1974).

[61] G.S. Hammond, J. Amer. Chem. Soc., 77, 334 (1955).

[62] E.R. Thornton, J. Amer. Chem. Soc., 89, 2915 (1967).
[63] J.C. Harris and J.L. Kurz, J. Amer. Chem. Soc., 92, 349 (1970).
[64] C. Minot and N. Trong Anh, Tetrah. Let., 45, 3905 (1975) ; see
     also C. Minot, Thèse de Doctorat d'Etat, Orsay, 1977.

Note added in proof

Some other papers dealing with the SN_2 reactions have been published recently. Olmstead and Brauman [65] have carried out an extensive gas-phase study and analyzed the nucleophilicity, the leaving group ability, the solvent effects (by comparing some selected reactions in solvents and in the gas-phase). From their results they also deduced that the energy profile along the reaction path was qualitatively similar to the ones obtained from the SCF calculations (Fig. 2 and 4). Two calculations of the reaction path for the reaction (11) [66] and the reaction $H^- + CH_3Cl \longrightarrow CH_4 + Cl^-$ [67] using the FSGO method and the LCAO-MO-SCF method (at the STO-3G and STO-31G levels) respectively failed to give even a local barrier, at variance with more refined calculations. They also point out to a rather early departure of the leaving Cl^- anion. Such a feature (as the late departure observed for the leaving H^- and F^- anions [39]) is probably an artefact of the method used for the determination of the reaction path [68] and would not occur on the so-called "intrinsic" reaction pathway which has been determined for the reaction (13) [68]. This reaction and the reaction (14) have also been analyzed through an energy decomposition scheme [69].

[65] W.N. Olmstead and J.I. Brauman, J. Am. Chem. Soc., 99, 4219 (1977).
[66] G. Simons and E.R. Talaty, Chem. Phys. Let., 56, 554 (1978).
[67] J.J. Woods, E.R. Talaty and G. Simons, to be published.
[68] K. Ishida, K.Morokuma and A. Komornicki, J. Chem. Phys., 66, 2153
     (1977).
[69] S. Nagase and K. Morokuma, J. Am. Chem. Soc., 100, 1666 (1978).

Table I. Ab initio calculations of potential energy surfaces for SN$_2$ reactions at a carbon center.

REACTION	METHOD	BASIS SET	REFERENCE
F$^-$ + CH$_3$F → FCH$_3$ + F$^-$	SCF	(9,5/4)[5,3/3]	27
H$^-$ + CH$_4$ → CH$_4$ + H$^-$	SCF	(9,3/3)[5,3/3]	26
H$^-$ + CH$_4$ → CH$_4$ + H$^-$	SCF	(11,7,1/6,1)[5,3,1/3,1]	35,38,39
F$^-$ + CHF$_3$ → CH$_3$F + F$^-$			
H$^-$ + CH$_3$F → CH$_3$F + H$^-$			
H$^-$ + CH$_3$F → CH$_4$ + F$^-$			
F$^-$ + CH$_3$F → CH$_3$F + F$^-$	SCF	(10,6,1/6,1)[5,3,1/3,1]	37,40
F$^-$ + CH$_3$CN → CH$_3$F + CN$^-$			
H$^-$ + CH$_4$ → CH$_4$ + H$^-$	CI	(11,7,1/6,1)[5,3,1/3,1]	41
F$^-$ + CH$_3$F → CH$_3$F + F$^-$			
Y$^-$ + CH$_3$X → YCH$_3$ + X$^-$	SCF,CI	(10,6,1/8,4,1/4,1)[6,3,1/4,2,1/2,1]	43
X,Y = H,F,Cl			
F$^-$ + CH$_3$F → CH$_3$F + F$^-$	SCF	4-31 G	45
C$_4$H$_8$ + H$^-$, C$_4$H$_8$ + F$^-$	SCF	STO 3G	44

Table II. Energy barrier calculated for the reaction $F^- + CH_3F \longrightarrow CH_3F + F^-$ as a function of the basis set used.

Calculated barrier [a] (kcal/mole)

Without d functions on C and diffuse p functions on F	With d functions on C and diffuse p functions on F	Reference
-8.8	-	27
-0.9	+7.9	35
-3.3	+7.2	37
-14.5	+2.1	42
-14.6	-	45
-	+4.2	43

[a] A negative value means that the energy at the mid-point of the reaction is lower than the energy of the reactants.

Table III. The change in correlation energy along the reaction path for a number of SN$_2$ reactions, based on CI calculations.

REACTION	CHANGE IN CORRELATION ENERGY (KCAL/MOLE)	REFERENCE
$H^- + H_2 \longrightarrow H_2 + H^-$	4.0	47
	4.4	41
	4.7–5.0	43
$H^- + CH_4 \longrightarrow CH_4 + H^-$	4.1	41
	15.2	48
	7.2	43
$F^- + CH_3F \longrightarrow CH_3F + F^-$	0.0	41
	4.6	43

Table IV. Calculated energy barrier <u>versus</u> activation energy in the gas-phase for a number of SN_2 reactions.

REACTION	CALCULATED BARRIER a,b (KCAL/MOLE)	ACTIVATION ENERGY a (KCAL/MOLE)
$H^- + CH_4 \longrightarrow CH_4 + H^-$	60. [39]	>6. [25]
$H^- + CH_3F \longrightarrow CH_4 + F^-$	3.8 [39]	3.6±0.1 [22]
$CN^- + CH_3F \longrightarrow CH_3CN + F^-$	22.6	>5. [22]
$Cl^- + CH_3Cl \longrightarrow CH_3Cl + Cl^-$	2.2 [43]	3.-4. [c]
$H^- + CH_3Cl \longrightarrow CH_4 + Cl^-$	-8.0 [43]	0.7±0.2 [22]
$F^- + CH_3Cl \longrightarrow CH_3F + Cl^-$	-15. [43]	0.1±0.2 [22]

a SCF values

b A negative value means that the energy of the "transition state" is lower than that of the reactants.

c Based on the data of Ref. 24.

Table V. The calculated SCF energy barriers for three series of SN$_2$ reactions with H$^-$, F$^-$ and Cl$^-$ as the leaving group.

REACTION	SCF ENERGY BARRIER (KCAL/MOLE)	NUCLEOPHILE, ELECTROPHILIC CENTER AND LEAVING GROUP
H$^-$ + CH$_4$ → CH$_4$ + H$^-$	60.[39]	H····C····H
H$^-$ + CH$_3$F → CH$_3$F + H$^-$	60.[39]	H····C····H
F$^-$ + CH$_4$ → CH$_3$F + H$^-$	72.[39]	F····C····H
Cl$^-$ + CH$_4$ → CH$_3$Cl + H$^-$	106.[43]	Cl····C····H
H$^-$ + CH$_3$F → CH$_4$ + F$^-$	3.8[39]	H····C····F
F$^-$ + CH$_3$F → CH$_3$F + F$^-$	7.9[39]	F····C····F
CN$^-$ + CH$_3$F → CH$_3$CN + F$^-$	23.[37]	CN····C····F
Cl$^-$ + CH$_3$F → CH$_3$Cl + F$^-$	60.[43]	Cl····C····F
H$^-$ + CH$_3$Cl → CH$_4$ + Cl$^-$	- 8.[43]	H····C····Cl
F$^-$ + CH$_3$Cl → CH$_3$F + Cl$^-$	- 15.[43]	F····C····Cl
Cl$^-$ + CH$_3$Cl → CH$_3$Cl + Cl$^-$	2.2[43]	Cl····C····Cl

Table VI. The C-H and C-F bond lengths corresponding to the leaving group for the transition states $(CH_4X)^-$ and $(FCH_3X)^-$ of reactions (23) and (24).

REACTION	C-H OR C-F BOND LENGTH (IN Å) IN THE TRANSITION STATE	REFERENCE
$H^- + CH_4 \longrightarrow CH_4 + H^-$	$H\cdots C\cdots H$ 1.74	39
$F^- + CH_4 \longrightarrow CH_3F + H^-$	$F\cdots C\cdots H$ 1.94	39
$Cl^- + CH_4 \longrightarrow CH_3Cl + H^-$	$Cl\cdots C\cdots H$ 2.12	43
$H^- + CH_3F \longrightarrow CH_4 + F^-$	$H\cdots C\cdots F$ 1.96	39
$F^- + CH_3F \longrightarrow CH_3F + F^-$	$F\cdots C\cdots F$ 1.88	39
$Cl^- + CH_3F \longrightarrow CH_3Cl + F^-$	$Cl\cdots C\cdots F$ 2.18	43

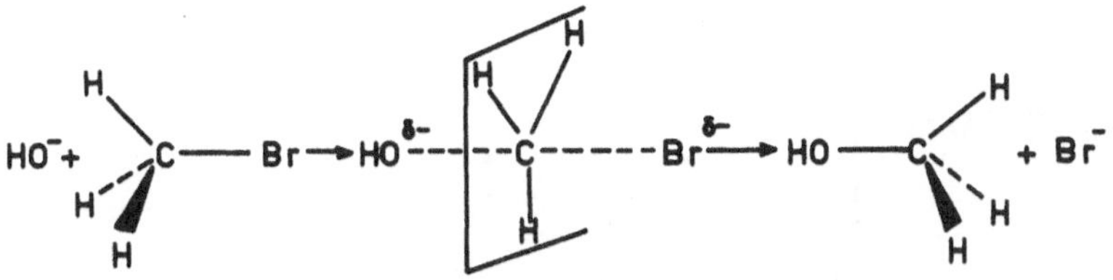

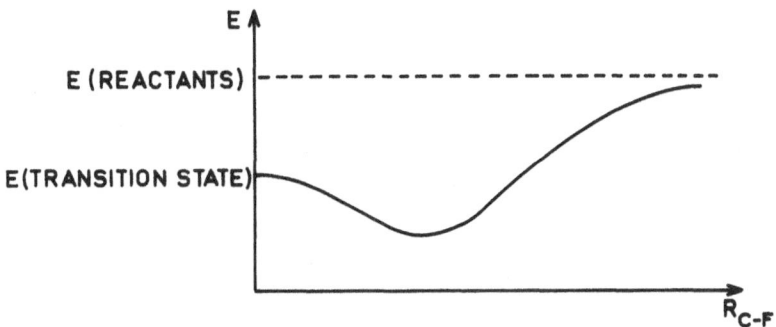

Fig. 1 The SN$_2$ bimolecular substitution and the postulated transition
        state.

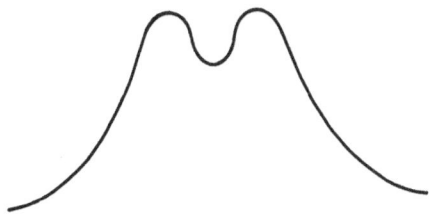

Fig. 2 Energy profile along the path of the reaction
        F$^-$ + CH$_3$F $\longrightarrow$ CH$_3$F + F$^-$ according to the SCF calculation of Ref. [27].

Fig. 3 Energy profile along the path of the reaction
        F$^-$ + CH$_3$F $\longrightarrow$ CH$_3$F + F$^-$ according to Ref. [32].

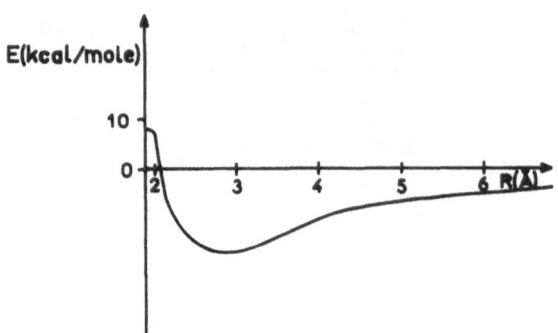

Fig. 4 Energy profile along the path of the reaction
        $F^- + CH_3F \longrightarrow CH_3F + F^-$ according to the SCF calculation of Ref. [39].

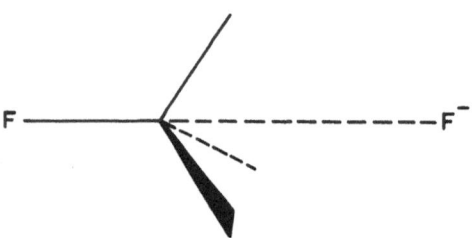

Fig. 5 The cluster $(CH_3F)F^-$ of C_{3v} symmetry.

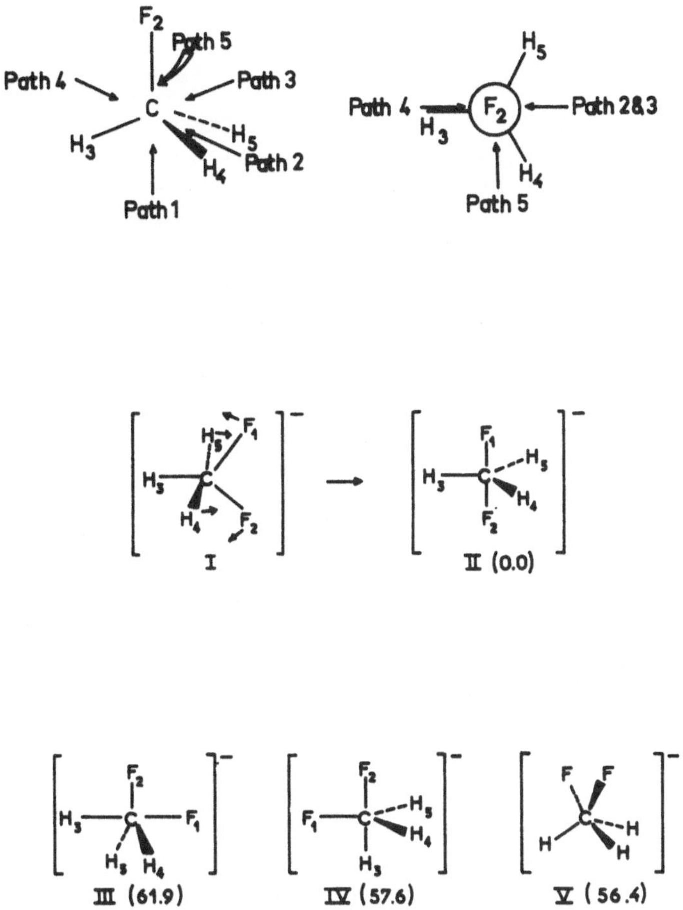

Fig. 6 The possible paths of attack of fluoride ion on methyl fluoride (side view and top view) and the relative stabilities (in kcal/mole) of the corresponding transition states according to Ref. [45].

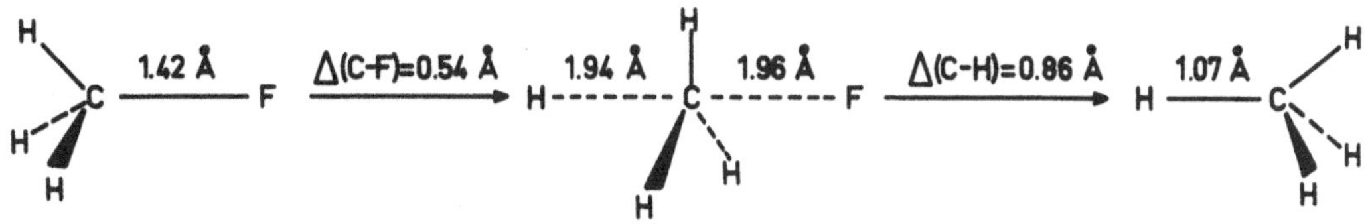

Fig. 7 The changes in bond length from the reactant to the transition state and the product for the reaction H$^-$ + CH$_3$F $\longrightarrow$ CH$_4$ + F$^-$ (from Ref. [39]).

CONTRIBUTION TO THE THEORETICAL STUDY OF REACTION MECHANISMS

G.Leroy, M.Sana, L.A.Burke and M.-T.Nguyen

Université Catholique de Louvain
Laboratoire de Chimie Quantique,
Bâtiment Lavoisier, Place Louis Pasteur, 1
1348 Louvain-la-Neuve   (Belgium)

ABSTRACT

We present the principal results that we have obtained in the theoretical study of the mechanisms of 1,3-dipolar cycloadditions, ring-
chain isomerizations and (2 + 2) and (2 + 4) cycloadditions.
The originality of our work rests essentially in the detailed analysis of the energetic, geometric, and electronic characteristics along
the reaction pathways. In particular, the reorganization of chemical
bonds is followed with the help of the evolution of charge centroids
for localized orbitals on the reaction path of the supersystem.

1. INTRODUCTION

This work constitutes a contribution to the theoretical study of
reaction mechanisms within the now classic framework of the supermolecule. We have chosen certain cycloaddition reactions, where we always
limit ourselves to the concerted approach of the partners, and certain
cyclization reactions.
In these reactions, called molecular, no rupture of a bond is produced such that the number of pairs of electrons is preserved. Generally, the n or σ doublets in the reactants become σ, n, or π doublets in
the reaction product. We have been able in this case to use the LCAO-
SCF-MO method without having to apply systematically a configuration
interaction. Furthermore, we have calculated only that portion of the
potential energy hypersurface for each supersystem which would be necessary for determining the path of the corresponding reaction.
The originality of our work lies in the detailed analysis of the
energetic, geometric, and electronic characteristics of the reaction
paths. In particular, the localization of the molecular orbitals along
the reaction path of each supersystem permits one to demonstrate the
reorganization of the chemical bonds during the transformation being
studied.

R. Daudel, A. Pullman, L. Salem, and A. Veillard (eds.), Quantum Theory of Chemical Reactions, Volume I, 91-144.
Copyright © 1979 by D. Reidel Publishing Company.

The ultimate aim of this work, which is still in its preliminary phase, is to accumulate a sufficient body of reliable theoretical results so as to be able to discover certain general laws concerning the mechanisms of the chosen reactions. We hope then to formulate qualitative models which would be more precise than those of resonance theory. These would permit one to interpret and predict the reactivity of compounds as well as to analyse in a critical way the empirical rules such as Hammond's postulat and least motion pathways.

2. METHODS AND PROGRAMS

The calculations have been carried out by the ab initio LCAO-SCF-MO method, using the GAUSSIAN-70 series of computer programs[1]. The STO-3G minimal basis set of contracted gaussians was used for the geometry optimizations. This basis set has been shown to give satisfactory geometries for a wide selection of organic compounds[2]. Nevertheless, several points on the potential energy hypersurfaces were recalculated using other basis sets, particularly the slightly more extended 4-31G basis set, which gives much better total energies for the compounds[3].

In addition, localized molecular orbitals (LMO) were obtained employing the Boys procedure[4] adapted to the GAUSSIAN-70 program[5]. With this procedure, the positions of the LMO charge centroids can be calculated. These centroids clearly define the centers of gravity of the electron clouds associated with those regions of a molecule which in the past have been intuitively assigned as core, bonding and lone pair regions. Furthermore, the change in the positions of the centroids along a reaction path can be used to describe the reorganization of the bonds during the corresponding chemical reaction.

In some cases, we have calculated the electrostatic potential[6] of the supermolecule using the ELPOT program[7].

Finally, in order to test our SCF results, we have performed different types of limited configuration interactions (CI) employing a computer program written by Segal's group[8]. The configurations were constructed from the mono- and diexcitations from the highest occupied molecular orbital (HOMO) to the lowest unoccupied (LUMO) (3 configurations), from the 2 π HOMO's to the 2 π LUMO's (15 configurations), and from the 3 HOMO's to the 3 LUMO's (55 configurations) of the supermolecule.

3. THE THEORETICAL APPROACH TO SEVERAL ADDITION REACTIONS

3.1. The 1,3-dipolar cycloadditions.

1,3-dipolar cycloadditions are those reactions with two centers on three center molecules leading to the formation of five membered heterocycles[9]. They bring into play a "dipole" possessing four π electrons delocalized on three centers and a "dipolarophile" containing a double bond, which can be activated or not. In figure 1. We give the dipoles and dipolarophiles which are presented in this work.

There has been much controversy by experimentalists over the mechanisms of these (2 + 4) electron reactions. Firestone[10] and Huisgen [11] have respectively proposed a two-step mechanism calling for a diradical intermediate and a concerted mechanism in which the reactants approach each other in parallel planes. A weak influence of the solvent polarity on the reaction rate, the stereoselectivity of the addition, and the strongly negative activation entropy all favor a concerted reaction.

In the work presented here we have only explored that portion of the potential energy hypersurface which corresponds to the concerted approach suggested by the experimental facts. Our results thus do not permit one to choose between the two different mechanisms but do lead to a detailed description of the concerted approach.

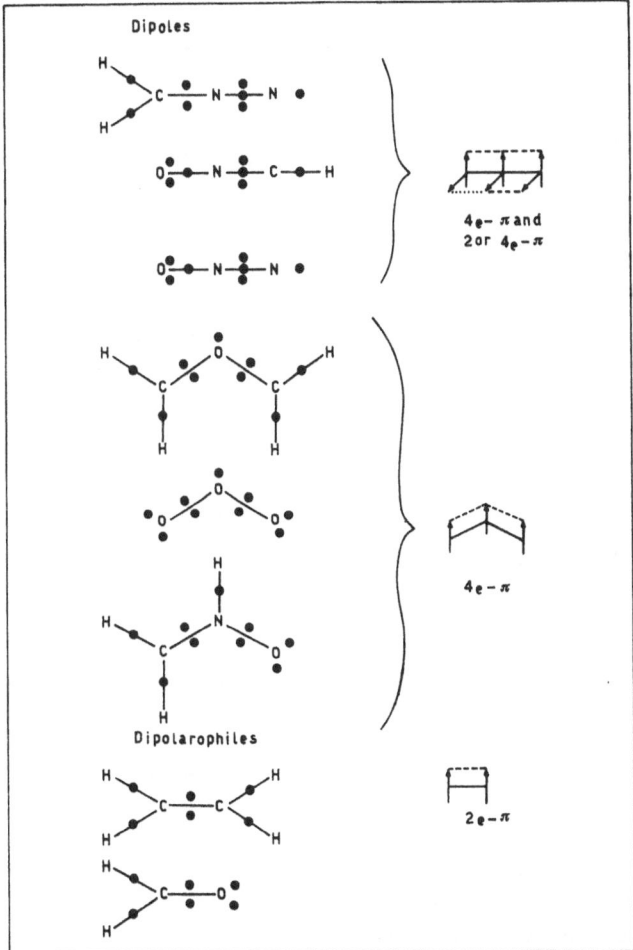

Figure 1. Examples of dipoles and dipolarophiles.

At this point in our research, we have studied the addition of ethylene with and without substituents successively on linear dipoles : diazoalkanes, nitrous oxide, fulminic acid[12], then on bent dipoles : carbonyl ylide, ozone, nitrone. Lastly, we studied the addition of formaldehyde to carbonyl ylide. These different reactions are given schematically in figure 2.

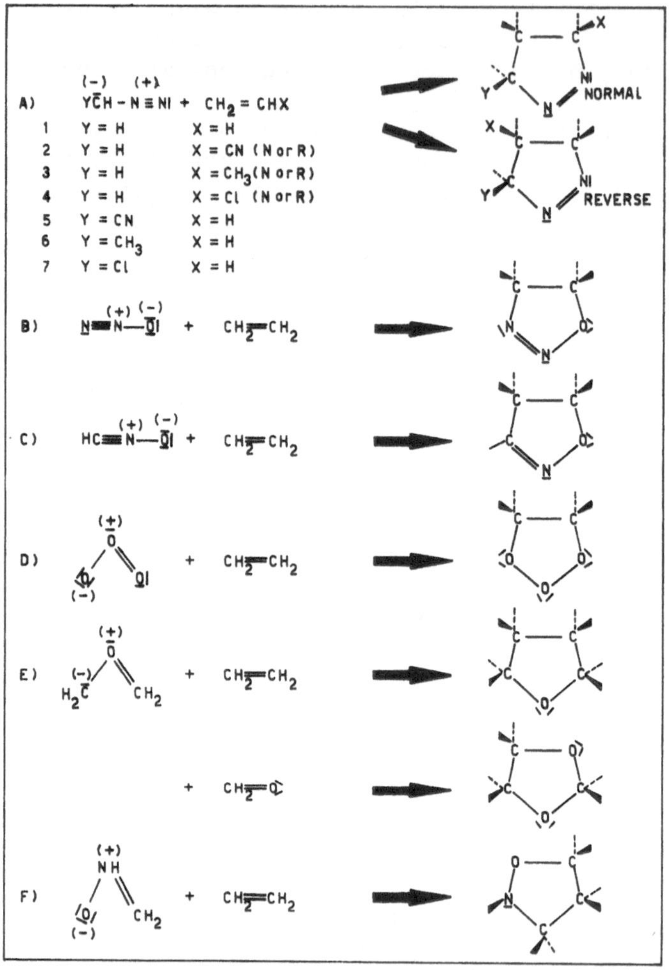

Figure 2. 1,3-dipolar cycloadditions.

We have calculated the total energies of the supersystems in function of the structural parameters, the details of which are given in figure 3 and whose symbols are the following :

    R   : distance between reactants ;
    τ : envelope angle of the heavy atoms ;
    β : angle between bonds 3-4 and 4-5 ;
    γ : repulsion angle of the substituents from the dipole ;
    δ : repulsion angle of the substituents from the dipolarophile ;
    α : angle of the rise of the N_4H bond (reaction with nitrone) ;

ν : angle C_3N_4H (reaction with nitrone) ;
θ : HCH or HCY angle in dipole ;
φ : HCH or HCX angle in dipolarophile ;
r_{12} : C_1C_2 distance in dipolarophile ;
r_{34} : distance between atoms 3 and 4 in dipole ;
r_{45} : distance between atoms 4 and 5 in dipole ;
D : lateral displacement of dipolarophile ;
ζ : inclination angle of the dipolarophile on the dipole.

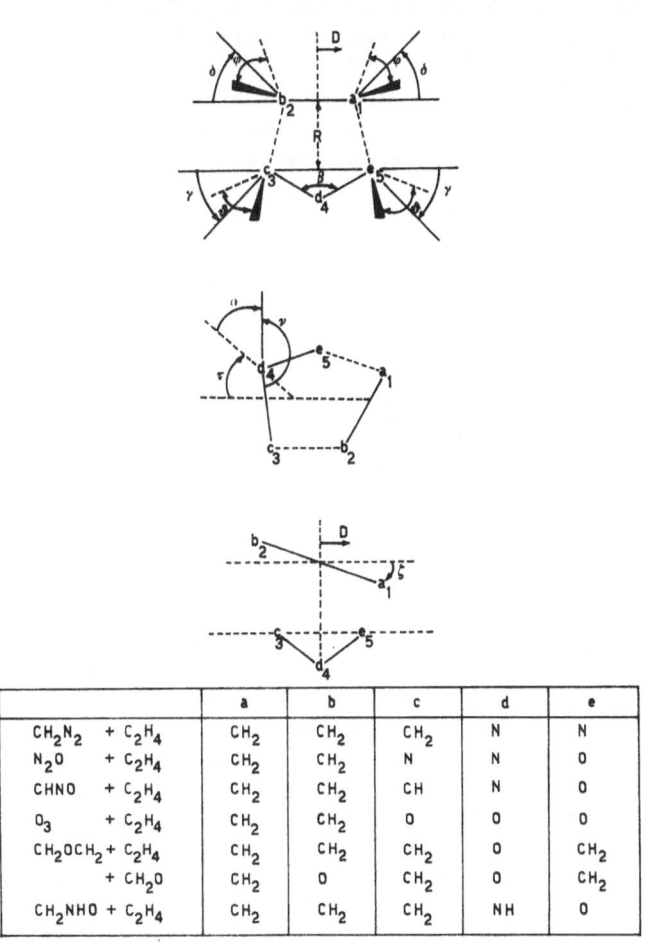

	a	b	c	d	e
CH_2N_2 + C_2H_4	CH_2	CH_2	CH_2	N	N
N_2O + C_2H_4	CH_2	CH_2	N	N	O
$CHNO$ + C_2H_4	CH_2	CH_2	CH	N	O
O_3 + C_2H_4	CH_2	CH_2	O	O	O
CH_2OCH_2 + C_2H_4	CH_2	CH_2	CH_2	O	CH_2
+ CH_2O	CH_2	O	CH_2	O	CH_2
CH_2NHO + C_2H_4	CH_2	CH_2	CH_2	NH	O

Figure 3. Geometrical parameters in the 1,3-dipolar cycloadditions.

We will describe the principal results obtained in our theoretical study in the following sections.

3.1.1. Reactions taking place with linear dipoles.

 A) The addition of diazoalkanes to olefins[13].

 A priori, the reactants can come together in approaches which keep the heavy atoms of the molecules either in parallel planes (P approach,

$\tau = 90°$) or in the same plane (C approach, $\tau = 0°$). In every case considered here, the calculations have shown that the C approach is favored
at distances smaller than 3 Å. We will successively analyze our results
in energetic, geometric, and electronic terms.

We present in table 1 the activation barriers and the reaction
energies calculated in the different basis sets for all the additions
of the diazoalkanes on olefins that were considered in this work. As
the theoretical reaction energies are generally overestimated, we have
evaluated the standard heats of reaction from the heats of formation of
the reactants and products, respectively, calculated by the partitioning
techniques of Allen[14] and of Benson[15]. The results that were obtained
are presented in the last column of table 1.

To our knowledge, only the activation energy for the diazomethane-
ethylene reaction has been experimentally determined. Huisgen proposed
the value of 14.9 kcal/mole[16] which is of the same order as that calculated with the 7s-3p basis set (contracted to 2s-1p). It is well
known that a quantitative agreement between the theoretical activation
barrier and the experimental activation energy can only be fortuitous
for these two concepts do not have entirely the same significance.
Furthermore, the theoretical activation barrier is very sensible to the
size of the basis set.

One can show that, whatever be the size of the basis set, the normal type addition of diazomethane to substituted ethylene is always
lower by several kcal/mole than that of the reverse type addition. In
generalizing we thus find again the empirical rule of von Auwers which
we express in these terms :
"In cycloadditions of diazomethane to monosubstituted ethylene the nor-
mal direction of the addition is that where the terminal nitrogen of
the dipole joins to the substituent bearing carbon of the dipolarophile".

One will also note that electron withdrawing substituents activate the ethylene and electron donators diminish its reactivity. On the
other hand, electron withdrawing groups desactivate diazomethane and
donor activate it. These results can easily be interpreted in light of
the direction of charge transfer – from the dipole to the dipolarophile
– in a reference reaction (vide infra). Chlorine, it is to be remarked,
which has the effects M– M+, modifies only slightly the reactivity of
ethylene whereas it activates diazomethane to the same degree as methyl.

If we now consider the geometric point of view, we will reiterate
that the approach of the reactants takes place (for R < 3.0 Å) with the
heavy atoms of the future ring in the same plane. Let us point out here
that the precision obtained for the distances is estimated to ± 0.03 Å
and for the angles ± 5°. The dipole progressively adopts a bent configuration and the hydrogens (or substituents) enter into positions that
prefigure those that they occupy in the final product (a 1-pyrazoline).

Within the precision of our calculations, all the activated complexes have practically the same geometry. Only the parameter D, which
corresponds to a lateral displacement of the dipole with respect to the
dipolarophile, varies a little from one reaction to another. On the whole, the activated complexes still resemble quite a bit the starting molecules as it is shown in table 2 where we compare the geometric parameters of the actived complexes to those of the reactants in the addition of
diazomethane to ethylene and to acrylonitrile (in the normal approach).

Table 1. Activation barriers and energies of reactions (kcal/mole).

Dipole	Dipolarophile	$\Delta E^{\ddagger}$		ΔE_0		ΔH_0°
		STO-3G	7s-3p	STO-3G	7s-3p	
Diazomethane	Ethylene	21.2	19.4	-111.6	-82.5	-27.0
	Acrylonitrile (N)	19.2	16.9	-101.3	-69.3	-18.5
	Acrylonitrile (I)	20.9	18.1	-102.8	-71.6	-20.2
	Propene (N)	24.0	22.4	-110.3	-77.4	-20.8
	Propene (I)	29.1	24.3	-109.8	-77.8	-19.7
	Vinyl chloride (N)	21.7	(20.0)[a]	-106.4	(-78.0)[a]	-16.3
	Vinyl chloride (I)	24.3	(22.7)[a]	-105.6	(-77.0)[a]	-24.4
Cyanodiazomethane	Ethylene	24.7	21.7	-101.0	-66.8	-25.4
Diazoethane		21.1	17.6	-108.5	-78.5	-16.6
Chlorodiazoethane		21.1	(17.6)[a]	-111.4	(-82,0)[a]	- 8.5

a Interpolated value

Table 2. Geometric parameters of two supersystems for three points on the reactions paths (Å or degrees).

	R	β	γ	δ	θ	φ	φ'	D	r_{34}	r_{45}	r_{12}	CC($\equiv$N)
Diazomethane + ethylene												
Reactants	∞	180	0	0	126	117	117	0.00	1.30	1.14	1.34	–
Transition state	2.25	150	10	10	120	114	114	0.00	1.30	1.14	1.34	–
Product	1.53	112	54	54	109	109	109	0.00	1.48	1.24	1.53	–
Diazomethane + acrylonitrile (N)												
Reactants	∞	180	0	0	126	117	116	0.00	1.30	1.14	1.34	1.42
Transition state	2.25	150	10	10	120	114	113	-0.03	1.30	1.14	1.34	1.45
Product	1.53	112	54	54	109	109	109	0.00	1.48	1.24	1.53	1.46

The populations of the bonds in formation permit one to calculate the asynchronism of a reaction. In this aim, one defines the rate of evolution of a bond at any point on the reaction path by the relation :

$$T_{AB}(R) = 100 \; \frac{E_{AB}(R) - E_{AB}(\infty)}{E_{AB}(\text{product}) - E_{AB}(\infty)} \tag{1}$$

where E_{AB} is the energy of the bond AB calculated with the corresponding population P_{AB} by the formula :

$$E_{AB} = a_{AB}P_{AB}^3 + b_{AB}P_{AB}^2 + c_{AB}P_{AB} \; (\text{kcal/mole})^{(13)} \tag{2}$$

The asynchronism is obtained by taking the weighted difference of the rates of evolution of the two bonds being formed in the activated complex :

$$A(AB/CD) = 100 \; \frac{T_{AB}(R^{\neq}) - T_{CD}(R^{\neq})}{T_{AB}(R^{\neq}) + T_{CD}(R^{\neq})} \tag{3}$$

So the value of the asynchronism always is comprised between -100% and $+100\%$; it becomes zero when the two new bonds have the same rate of evolution at the transition state. In each reaction studied the asynchronism is rather small ($7 \leqslant A \leqslant 22$) and in favor of the carbon-carbon bond.

Overall, the cycloadditions of diazoalkanes to olefins in their concerted channels are practically synchronous reactions, passing by a planar activated complex, and geometrically similar to the reactant molecules.

The electronic reorganizations which are produced along the reaction pathway can be followed thanks to atomic populations and to the centroids of charge for localized molecular orbitals (LMO). Thus, the charge transfer of one reactant towards another is estimated by the relation :

$$t = - \sum_A q_A \; , \tag{4}$$

where q_A designates the net charge and the sum is taken over all the atoms of the dipolarophile. The transfer is then positive when it takes place from the diazoalkane towards the olefin.

With the cycloadditions of the diazoalkanes studied here the charge transfer is always positive. It varies between 0.06 for ethylene and around 0.10 for acrylonitrile, at the transition state, where it attains its maximal value. In accord with qualitative expectations, one finds that the transfer increases in the presence of an electron withdrawing group on the dipolarophile or a donor on the dipole and inversely, diminishes when the groups on the reactants exert the opposite effects.

The reorganization of the chemical bonds during the course of a reaction can be qualitatively described by the charge centroids of LMO's.

As an example, we show in figure 4 the evolution of the centroids of charge along the path of the cycloaddition of diazomethane to ethylene[17].

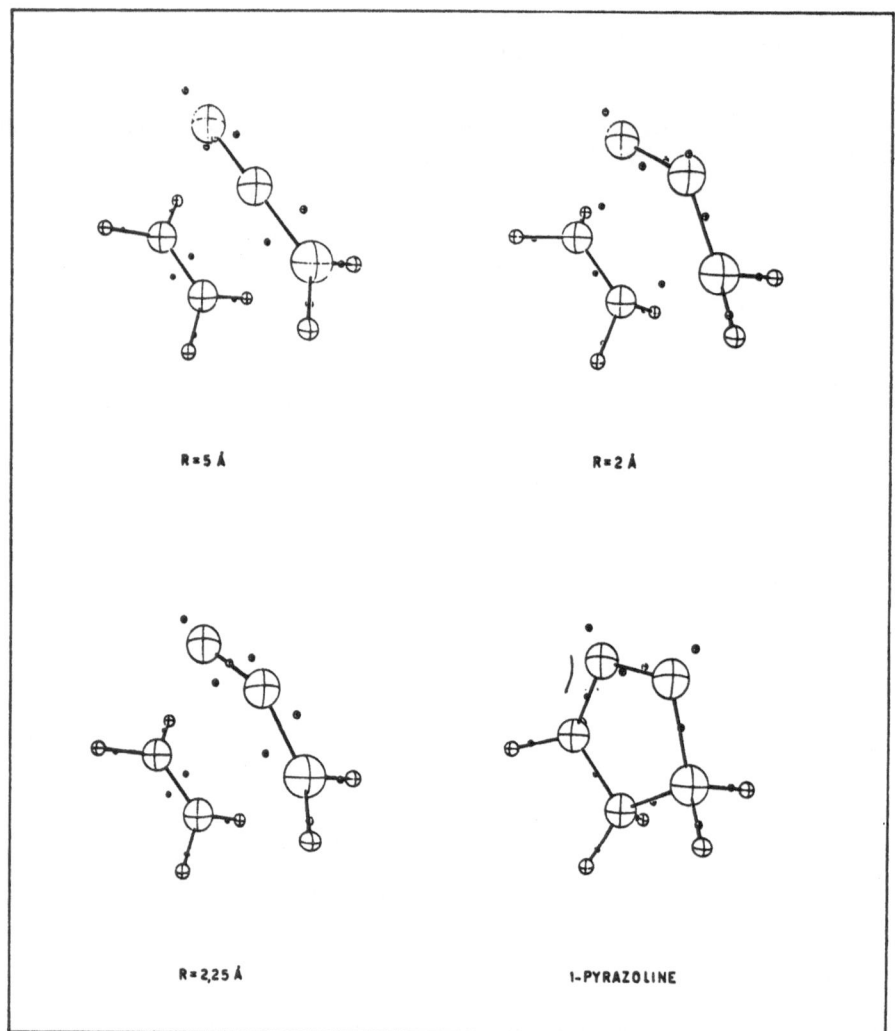

Figure 4. Evolution of the centroids of charge along the path of the cycloaddition of diazomethane to ethylene.

One finds that at the transition state the bonds between the partners are not yet formed, even though there is some displacement of the centroids. When the transition state is well past, one observes a complete reorganization of the bonds which become identical, in number and kind, to those in the product 1-pyrazoline. One remarks that the electrons of the bond C_2C_3 come from the dipole and those of bond C_1N_5 from the dipolarophile. Finally, the lone pair on N_4 is the result of the migration of a pair of π electrons from the triple bond N_4N_5.

The addition of diazomethane to ethylene is thus carried out by a cyclic displacement of the π electrons of the partners and at a point posterieur to the transition state.

It is reasonable to admit that the direction of the displacement of the centroids that is observed in the reference reaction – ethylene plus diazomethane – is the "natural" direction of the formation of the bonds in the 1,3-dipolar cycloadditions of diazoalkanes to olefins. With this point of view the effects of the substituents in the normal additions can be interpreted as follows : activating groups favor the rotation of the bonds in this "natural" direction and desactivating groups oppose it. Likewise, one can understand how in the reverse addition of diazomethane to acrylonitrile the CN group induces an electronic movement in the opposite direction to the "natural" one, thus causing a raising of the activation energy. These different effects are illustrated schematically in figure 5.

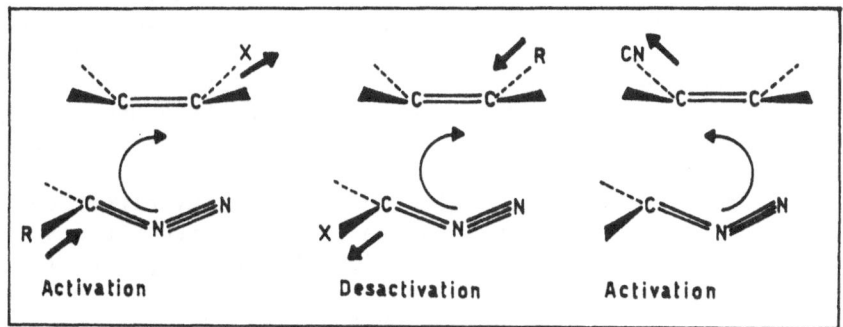

Figure 5. Electronic effects of the substituents in the cycloadditions of diazoalkanes to substituted ethylenes.

B) The addition of nitrous oxide to ethylene[18].

Our calculations have shown that for distances equal or inferior to 3 Å (R ⩽ 3 Å) the five heavy atoms of the reactants remain in the same plane. As in the case of diazomethane, this type of approach is labelled "C" (τ = 0°).

With the STO-3G basis set the activation barrier and the energy of reaction equal 21.4 and – 93.9 kcal/mole, respectively. The structural parameters that were obtained after optimization are presented in table 3.

In order to assure ourselves of the quality of the transition state obtained we have recalculated all the points near the transition state with the limited configuration interactions (CI) employing the mono and diexcitations between the following groups of orbitals :

- the highest occupied molecular orbital (HOMO) and lowest unoccupied molecular orbital (LUMO) of the supermolecule ;
- the two HOMO's and two LUMO's ;
- the three HOMO's and three LUMO's.

Table 3. Geometric parameters along the reaction path for nitrous
         oxide to ethylene.

	R	τ	β	δ	ϕ	d	r_{12}	r_{34}	r_{45}
Reactants	∞	0	180	0	117	0.0	1.34	1.15	1.27
Transition state	2.15	0	144	10	114	0.1	1.34	1.17	1.30
Product	-	0	114	55	109	-	1.54	1.26	1.42

The wave functions thus obtained are linear combinations of determinants corresponding respectively to 3, 15 and 55 configurations. A priori,
the (3 x 3) CI should be insufficient since it only includes excitations
from the dipolarophile towards the dipole. On the other hand, the (55 x
55) CI possesses the minimal size required to include the excitations
between the frontier π molecular orbitals of the two reactants, i.e.
two occupied and two unoccupied MO's of the dipole (degenerate at R = ∞)
plus one occupied and one unoccupied MO of ethylene. The results obtained in SCF and the three CI's are given in table 4 where we give the
total energies calculated for different values of R and τ with all
other parameters optimized.

        The different CI's used here do not have any effect on the enve-
lope angle τ as the transition state in every case possesses the structure C. We note, however, that the (55 x 55) CI displaces the transition
state towards the product by a value of 0.05 Å, along the reaction coordinate R. Overall, the activation energies and heats of reactions
equal (in kcal/mole) :

        21.4   and   - 91.9   SCF
        19.5   and   -100.9   (3 x 3)   CI
        24.3   and   - 62.9   (15 x 15) CI
        31.3   and   - 37.2   (55 x 55) CI

In figure 6 we give the change in energy along the reaction path. This
diagram clearly shows that it is above all the reactants and then the
activated complex which are effected by the larger CI methods, and from
this fact the activation energies are raised and the heats of reaction
lowered passing from SCF up to a (55 x 55) CI. We have taken the supersystem at 7 Å as the reference point in the energy calculations of the
chosen CI's. At this distance the SCF energy of the supersystem equals
the sum of the SCF energies of the separated molecules. The square of
the coefficient of the SCF ground state configuration is 0.9 (the sum
of the squares of all the coefficients being 1.0) in the (55 x 55) CI and
the nuclear structures of the activated complexes are very nearly the
same for the SCF and CI methods. Then the system can be described by
properties related to SCF alone.

        Since the nuclear structures of the reactants and the activated
complex are rather close, it is reasonable that the electronic characteristics be little changed in this region of the hypersurface.

Table 4. Total energies in function of structural parameters R and τ (a.u.).

		R = 2.20 Å	R = 2.15 Å	R = 2.10 Å	R = 2.05 Å
τ = 90°	SCF	-258.19815	-258.18250	-258.16311	-258.14344
	3 x 3	-258.19816	-258.18252	-258.16315	-258.14359
	15 x 15	-258.23950	-258.22298	-258.20331	-258.18349
	55 x 55	-258.29935	-258.28167	-258.26089	-258.23947
τ = 45°	SCF	-258.23231	-258.22763	-258.22387	-258.22176
	3 x 3	-258.23270	-258.22842	-258.22557	-258.22478
	15 x 15	-258.28084	-258.27457	-258.26932	-258.25592
	55 x 55	-258.33252	-258.32539	-258.31958	-258.31549
τ = 0°	SCF	-258.24403	-258.24281	-258.24351	-258.24598
	3 x 3	-258.24572	-258.24566	-258.24929	-258.25452
	15 x 15	-258.30598	-258.30441	-258.30532	-258.30782
	55 x 55	-258.34369	-258.34007	-258.33880	-258.33931

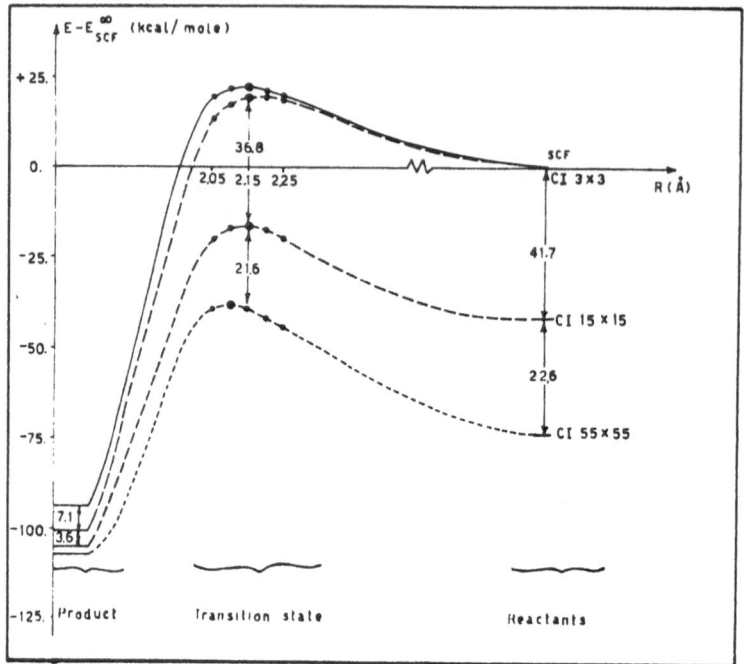

Figure 6. Relative energies along the reaction path for nitrous oxide
             to ethylene.

In table 5 we give the atomic populations and bond overlap populations
for three points on the SCF reaction path. As the overlap populations
bear out, it is only after the transition state where the major electronic reorganization takes place. In fact, the rate of formation of
the bonds (1) between the systems equals 6.7 % for bond C_2N_3 and 12.0 %
for C_1O_5, and the asynchronism (3) is 28.3 % in favor of C_1O_5. The
charge transfer (4) is some - 0.01 electrons and is towards the nitrous
oxide.
       With the help of the localization method of Foster and Boys one
can follow the evolution of the centroids of charge associated with
the localized molecular orbitals. The results obtained are represented
in figure 7. One observes the major reorganization of the bonds beyond
the transition state, towards the product. The intersystem σ bond C_1O_5
is formed by a transfer from bond N_4O_5, the lone pair on N_4 from bond
N_3N_4, and finally the σ bond C_2N_3 from the double bond C_1C_2 on ethylene . It is to be noted also that the evolution of the centroids for
bonds C_1O_5 and C_2N_3 illustrates the asynchronism that was first observed with the population analysis of Mulliken.

       C) The addition of fulminic acid to ethylene.

       The calculation of the potential energy hypersurface for this
reaction has been carried out by D. Poppinger[12]. We reproduce here
the essential results obtained by this author.

Table 5. Population analysis along the reaction path for nitrous oxide to ethylene (STO-3G basis set).

Atomic population

	C_1	$H(C_1)$	C_2	$H(C_2)$	N_3	N_4	O_5
Reactants	6.126	0.937	6.126	0.937	6.999	6.859	8.142
Transition state	6.107	0.929	6.157	0.935	6.983	6.904	8.121
Product	6.023	0.918	6.064	0.906	7.126	6.975	8.163

Bond overlap population

	C_1C_2	N_3N_4	N_4O_5	C_2N_3	C_1O_5
Reactants	0.599	0.556	0.226	0.000	0.000
Transition state	0.580	0.530	0.228	0.028	0.027
Product	0.351	0.412	0.207	0.285	0.251

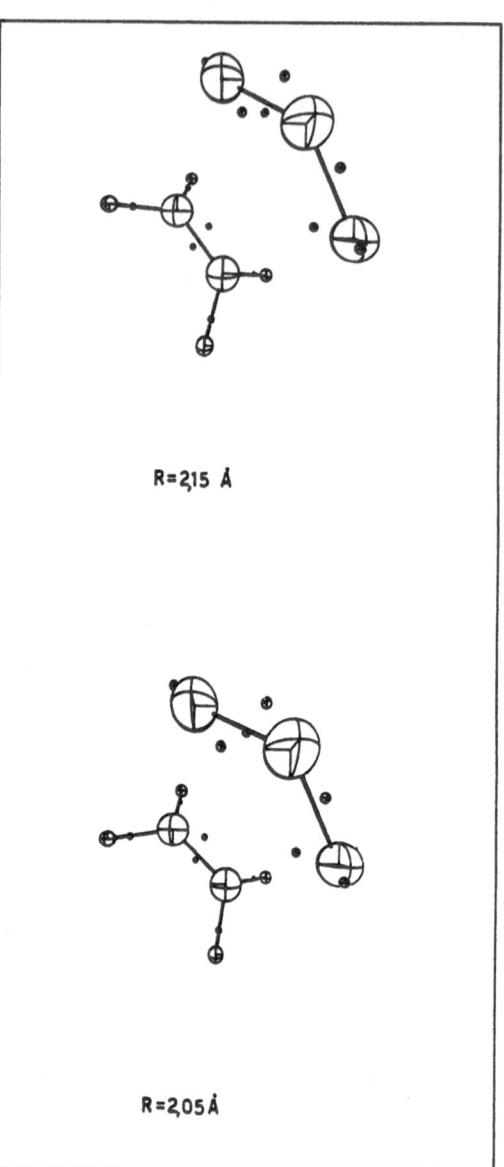

Figure 7. Evolution of charge centroids along the reaction
path for nitrous oxide
to ethylene.

We also indicate the values calculated with equation (1) to (4) as well
as the centroids of charge which we have found for the reported geometries. D.Poppinger finds a transition state where the five heavy atoms
are in the same plane ($\tau = 0°$). The structural parameters obtained are
given in table 6. With the STO-3G basis set the activation barrier
equals 15.5 kcal/mole and the heat of reaction is -132.8 kcal/mole,
while the transition state lies at 2.30 Å on the reaction coordinate R.
At the transition state the rate of formation of the intersystem bonds
C_1O_5 and C_2C_3 equal 10.1 % and 5.4 %, respectively. The asynchronism of
this reaction is 30.3 % in favour of the bond C_1O_5. The charge transfer
at this point is 0.02 electrons towards ethylene.

Table 6. Geometrical parameters along the reaction path for fulminic acid to ethylene (D.Poppinger).

	R	τ	β	δ	γ	ϕ	r_{12}	r_{34}	r_{54}
Reactants	∞	–	180	0	0	116	1.34	1.16	1.29
Transition state	2.25	0	144	10	35	115	1.34	1.19	1.29
Product	–	0	110	50	60	108	1.55	1.41	1.41

In figure 8, we present the evolution of the charge centroids obtained by the Foster-Boys method.

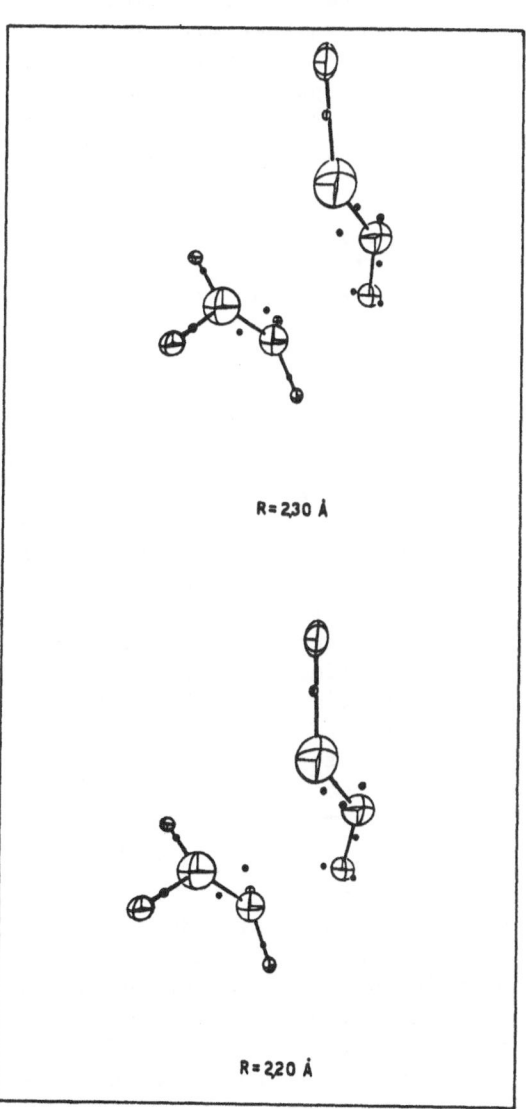

R = 2.30 Å

R = 2.20 Å

Figure 8. Evolution of the centroids along the reac-
tion path for fulminic
acid to ethylene.

One observes a rotation of these centroids leading to the formation of a lone pair on N_4 at the expense of the C_3N_4 π system and to the formation of two new σ bonds, C_1O_5 and C_2O_3 coming respectively from the π bonds N_4O_5 and C_1C_2.

3.1.2. Reactions taking place with bent dipoles.

A) The addition of ozone to ethylene[19].

For this reaction we will cite only the following points :

- at large distances (R = 5 Å) the approach of the reactants is carried out in two parallel planes. One thus finds the model proposed by Huisgen for 1,3-dipolar cycloadditions in general ;
- at the transition state (R = 2.30 Å) the partners are presented face to face with an envelope angle of 63° ; the geometry of the activated complex is thus intermediate between those resulting from approaches P and C ;
- the charge transfer is in the direction from the dipolarophile to the dipole with a value of − 0.02 electrons in the transition state ;
- with the STO-3G basis set, the activation barrier equals 1.3 kcal/mole and the heat of reaction is − 136.8 kcal/mole. By Benson's technique of groups one obtains an "experimental energy of reaction" of − 49 kcal/mole.

B) The addition of carbonyl ylide to ethylene and to formaldehyde.

The study of the theoretical cycloaddition of carbonyl ylide and ethylene furnishes the following informations[20] :

- as in the preceeding cases, the reactants approach each other in two parallel planes but the envelope angle remains 90° until the transition state. Huisgen's model is then entirely respected for this reaction ;
- after the transition state all the heavy atoms return to the same plane but then must leave this plane for the product, tetrahydrofuran whose envelope angle is 33° ;
- one finds the charge transfer to be 0.09 electron from the dipole to the dipolarophile at the transition state ;
- with the STO-3G basis set the activation barrier and the energy of reaction are equal respectively to 6.0 kcal/mole and − 181.2 kcal/mole.

We have also calculated the reaction path for the addition of carbonyl ylide to formaldehyde. This reaction no longer possesses the symmetry element of the cycloaddition of carbonyl ylide to ethylene. As such, one can expect a rotation of the centroids of charge and an asynchronism in the formation of the new bonds.

The values of the structural parameters along the STO-3G reaction path are given in table 7. They are quite close to those obtained after optimization in the preceding case[20].

Tableau 7. Geometrical parameters along the reaction path for carbonyl ylide to formaldehyde.

	R	τ	β	γ	δ	θ	ϕ	d	ζ	$r_{C_2O_1}$	$r_{C_3O_4}$
Reactants	∞	−	129	0	0	124	115	−	−	1.22	1.30
Transition state	2.15	90	121	20	20	120	112	−0.06	2	1.26	1.30
Product	−	40	108	20	59	108	108	−	−	1.43	1.43

The activated complex is, however, nearer to the product along the reaction coordinate. We have recalculated the potential energy hypersurface in the neighborhood of the transition state using several differently limited CI. The determinants employed include successively one, two, and three couples of HOMO's and LUMO's.

In every case, and within the limits of the precision obtained in our calculations, the CI does not modify the structural parameters and, in particular, the envelope angle τ from the SCF result. The position of the transition state is displaced $0.05\,\text{Å}$ towards the reactants by the (15×15) and (55×55) CI's and $0.05\,\text{Å}$ towards the product in the (3×3) CI . It is to be noted that in this supersystem the first HOMO and first LUMO are derived both from the dipole and so the interaction is expected to be much greater than in the case of nitrous oxide. Still, a (15×15) CI is expected to be the minimally accepted CI since it includes excitations from the π frontier orbitals of both reactants.

The activation barriers and the heats of reaction for this reaction equal respectively :

3.9 kcal/mole and − 171.9 kcal/mole in SCF
16.8 kcal/mole and − 131.9 kcal/mole in (3 x 3) CI
17.8 kcal/mole and − 104.9 kcal/mole in (15 x 15) CI and
17.7 kcal/mole and − 105.2 kcal/mole in (55 x 55) CI

In figure 9, we give the E(R) curve for the different calculations carried out. From these results and for a minimal basis set it would seem that the (15×15) CI is sufficient to describe the electronic evolution along this reaction path. Of the sum of the squares of the coefficients of these 15 configurations, the ground state SCF contributes overwhelmingly, being 87 % of the sum. Thus, a description of this reaction is possible using only the SCF function.

In table 8 we give the results of the population analysis for the three principal points on the reaction coordinate. At the transition state the rates of formation of the intersystem bonds are 15.3 % for bond C_2C_3 and 15.7 % for O_2C_5 , or an asynchronism of 1.3 % in favour of O_1C_5. At this point the transfer of charge is 0.163 electron towards formaldehyde.

The movements of the centroids of charge for several points along the reaction is represented in figure 10. We find that the intersystem σ bond O_1C_5 is formed at the expense of the double bond C_2O_1 of formaldehyde, that the σ bond C_2C_3 comes from an electron pair of bond C_3O_4

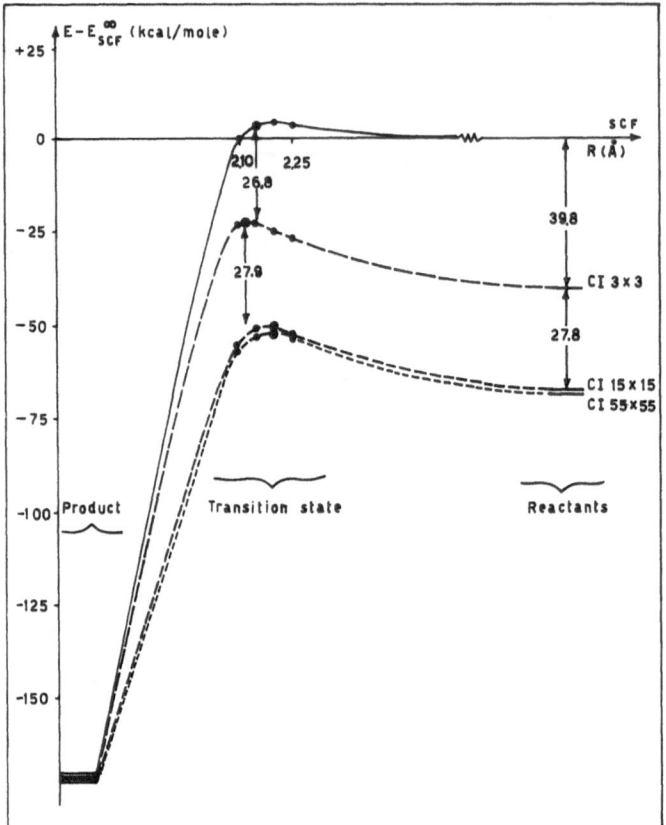

Figure 9. Relative energy for the reaction path for carbonyl ylide to
formaldehyde.

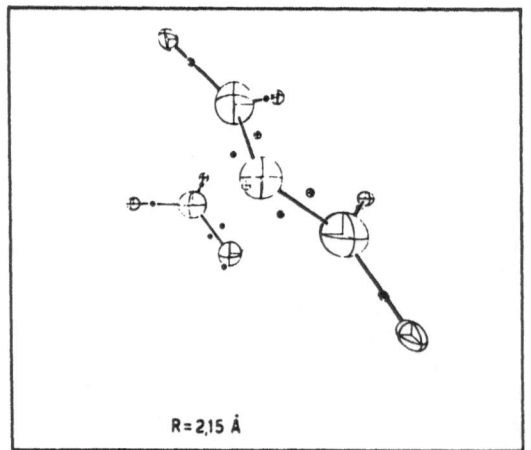 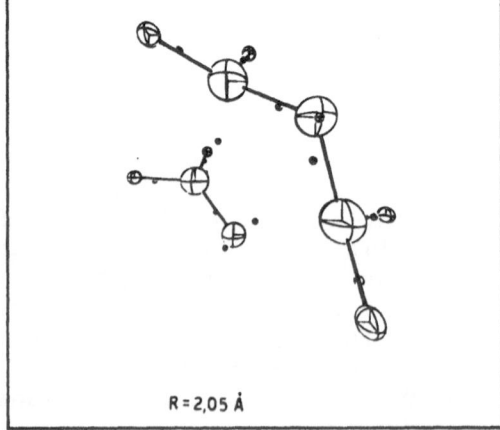

Figure 10. Evolution of charge centroids along the reaction path for
carbonyl ylide to formaldehyde.

Table 8. Population analysis along the reaction path for carbonyl ylide to formaldehyde.

Atomic population

	O_1	C_2	$H(C_2)$		C_3	$H(C_3)$		O_4	C_5	$H(C_5)$	
Reactants	8.186	5.930	0.942	0.942	6.065	0.948	0.934	8.105	6.065	0.948	0.934
Transition state	8.261	5.980	0.962	0.960	6.042	0.937	0.923	8.136	5.970	0.922	0.907
Product	8.251	66.012	0.936	0.931	6.009	0.963	0.919	8.256	5.892	0.842	0.916

Bond overlap population

	O_1C_2	C_3O_4	O_4C_5	C_2C_3	C_5O_1
Reactants	0.444	0.371	0.371	0.000	0.000
Transition state	0.402	0.354	0.373	0.069	0.037
Product	0.267	0.264	0.268	0.349	0.259

of the dipole and that the lone pair on O_4 is provided by the double bond O_4C_5. Of the two σ bonds in formation, the O_1C_5 new bond is the most precocious.

C) The addition of nitrone to ethylene[21].

The cycloaddition of nitrone to ethylene leads to isoxazolidine (figure 11).

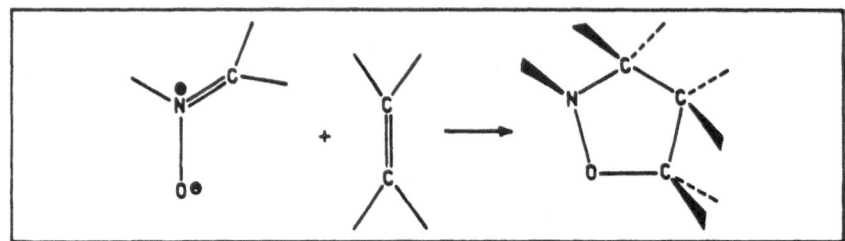

Figure 11. Cycloaddition of nitrone to ethylene.

We have completely optimized the geometry of nitrone in the STO-3G basis set, the results of which are given in table 9 along with other results taken from the literature. We have kept the experimental values[22] of the bond lengths and angles in isoxazolidine. Only the parameters α and τ have been optimized in this compounds. The absolute minimum is obtained for τ = 37° and α = 64°.

Table 9. Geometric parameters of nitrone.

Method	Compound	NO(Å)	CN(Å)	CH(Å)	NH(Å)	$\hat{CNO}$(°)	$\hat{HNC}$(°)	$\hat{HCH}$(°)
RX [23]	p-Cl-φ-CH=N(CH₃)-O	1.28	1.31	–	–	125	119	–
SMO [24]	φ-CH=N(CH₃)-O	1.33	1.32	–	–	120	–	–
INDO [25]	CH₂=N(OCH₃)-O	1.30	1.27	–	–	120	120	120
CNDO/2 [26]	CH₂=N(H)-O	1.21	1.32	1.08	–	120	120	–
MINDO/3 [27]	CH₂=N(H)-O	1.23	1.27	1.10	1.06	133	110	115
STO-3G [*]	CH₂=N(H)-O	1.32	1.31	1.08	1.05	129	114	119

(*) This work

In figure 12, we give the potential energy curves corresponding to the inversion on nitrogen calculated for several values of τ. We find that the pseudoaxial (p.a.) form is more stable by 3.1 kcal/mole than the pseudoequatorial (p.e.). The barrier to nitrogen inversion is 28 kcal/mole for τ = 37° (p.a. (τ = 37°, α = 64°)) → p.e. (τ = 37°, α = - 66°)).

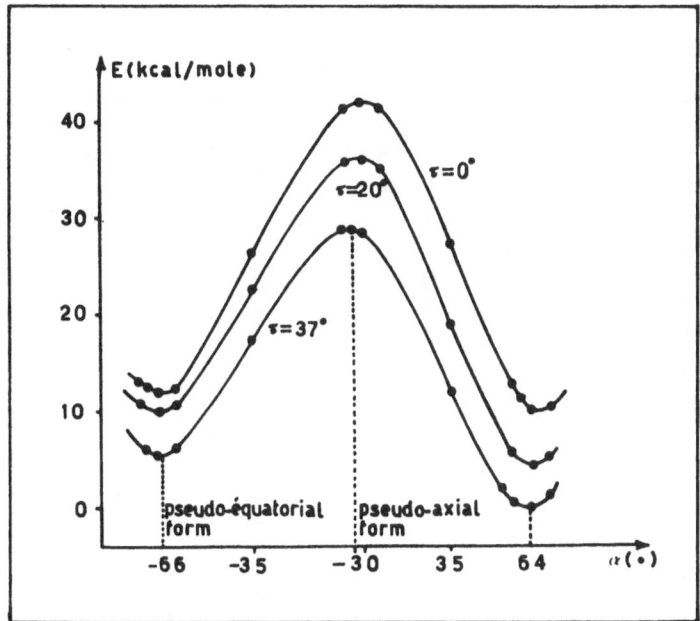

Figure 12. Inversion barrier of the nitrogen atom in isoxazolidine.

Also, the necessary energy for the inversion of the ring envelope is only 10.3 kcal/mole (p.a. ($\tau = 37°$, $\alpha = 64°$) → p.e. ($\tau = 37°$, $\alpha = 64°$)). The passage from the p.a. form to the p.e. form takes place then preferentially by an inversion of the isoxazolidine envelope. The geometric parameters of the most stable conformation of isoxazolidine are presented in figure 13.

Figure 13. Geometrical parameters in isoxazolidine.

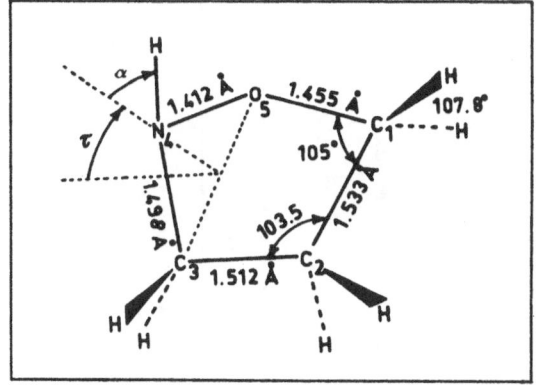

For the calculation of the potential energy hypersurface the parameter R is again chosen as the reaction coordinate. For each value of R the 12 other parameters were optimized. The energetic minima and the values of their corresponding parameters are presented in table 10.

It is interesting to analyse the evolution of the optimum values of the angle τ.

Table 10. Geometric parameters and energies (STO–3G) along the nitrone–ethylene reaction path.

R(Å)	τ(°)	β(°)	γ(°)	δ(°)	θ(°)	φ(°)	ν(°)	α(°)	r_{12}(Å)	r_{34}(Å)	r_{45}(Å)	D(Å)	E(a.u.)
∞	90	129	0	0	119	117	114	0	1.34	1.31	1.32	0.00	-243.638912
5.00	90	129	0	0	119	117	114	0	1.34	1.31	1.32	0.00	-243.638755
3.00	90	129	0	0	119	117	114	0	1.34	1.31	1.32	0.32	-243.636063
2.50	90	127	5	5	119	117	114	0	1.34	1.31	1.32	0.31	-243.620425
2.25#	90	125	10	10	119	117	115	3	1.34	1.31	1.32	0.19	-243.597075
2.15	33	122	15	25	117	115	119	50	1.41	1.40	1.36	-0.02	-243.602833
2.08	32	119	15	30	116	114	119	50	1.46	1.42	1.37	-0.03	-243.613265
2.00	25	116	15	35	114	113	120	60	1.49	1.46	1.39	-0.03	-243.637190
1.70	10	115	15	42	113	112	123	63	1.51	1.48	1.40	-0.03	-243.766704
1.48*	37	102	53	52	108	108	106	64	1.53	1.50	1.41	-0.02	-243.858363

(#) Transition state   (*) Isoxazolidine

One finds that at large distances the approach of the reactants is in two parallel planes ($\tau = 90°$, approach P); however, at shorter distances ($R < 2.25$ Å) the envelope angle is reduced gradually to 10° for $R = 1.7$ Å and then is increased to 37° in the reaction product. Thus, the approach C, where all heavy atoms lie in the same plane, is never realized.

We were interested in the barrier to the inversion of the nitrogen in the neighborhood of the transition state. The results that were obtained are presented in figure 14 which shows the behavior of the function $E(R, \alpha)$. One finds that at the transition state the curve presents a unique minimum for $\alpha = 3°$. It is only at a distance of 2.08 Å that one sees two minima appear, the lower corresponding to $\alpha = 50°$, the other at $\alpha = -50°$. So, very early after the transition state the hydrogen atom is placed in a pseudoaxial position. Grée, Tonnard and Carrié[28] have been lead to the same conclusion in their experimental results.

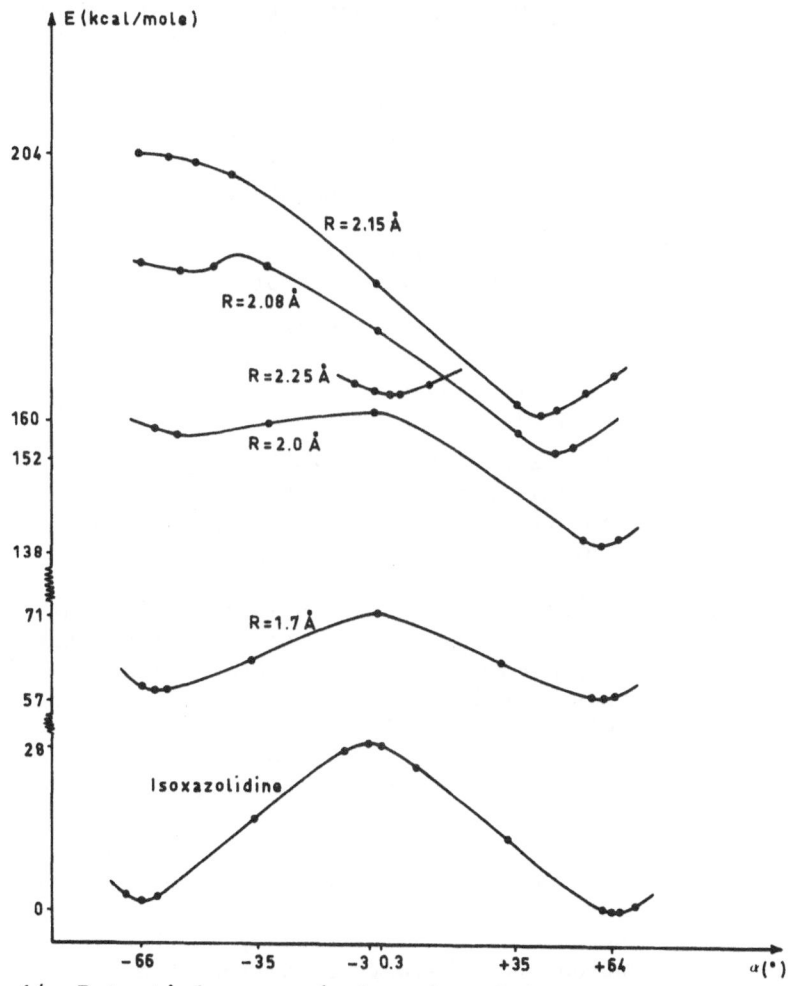

Figure 14. Potential energy in function of the parameters R and α.

With the STO-3G basis set the activation energy of the cycloaddition of nitrone with ethylene is 26.3 kcal/mole and the heat of reaction - 137.7 kcal/mole. We point out for a comparison that in the reaction of N-methyl, C-phenyl nitrone with an activated olefin the activation energy is between 16 and 18 kcal/mole according to the nature
of the substituent on the dipolarophile[29].

The results of the Mulliken population analysis showed an electron
transfer from the dipole towards the dipolarophile which is maximal in
the region of the transition state ($\sim$ 0.1). Furthermore, in the activated complex the rate of formation of bonds C_1O_5 and C_2C_3 are respectively 8.2 % and 3.7 %. This leads to an asynchronism of 37.8 % in favour of the CO bond.

The evolution of the chemical bonds in the supermolecule is described in figure 15 in terms of centroids of charge for several points
along the reaction path. One finds that until a distance of R = 2.25 Å
the evolution of the centroids is small. Thus, with the STO-3G basis
set the electronic structure of the activated complex is quite similar
to the reactants. As soon as this point on the reaction path is cros-
sed, there follows a complete reorganization of the bonds.

In order to locate better the reliability of the minimal basis set
results, we have recalculated several points of the reaction path with
the 4-31G basis set of Pople's group. We have also applied a configuration interaction to the minimal basis set.

We summarize here below the results for the 4-31G basis set.

$$E \text{ (nitrone)} = -168.552046 \text{ a.u.}$$
$$E \text{ (ethylene)} = -77.920910 \text{ a.u.}$$
$$E \text{ (transition state)} = -246.416358 \text{ a.u.}$$
$$E \text{ (isoxazolidine)} = -246.536098 \text{ a.u.}$$

The activation barrier is now 35.5 kcal/mole and the heat of reaction
- 39.6 kcal/mole. We recall that the only known activation energies
(16-18 kcal/mole) concern reactions between a highly substituted nitrone and activated olefins and are not then directly comparable to
our results.

Furthermore, the population analysis in the 4-31G basis set confirm the results obtained in STO-3G. At the transition state the charge transfer from the dipole to the dipolarophile equals 0.077 electron.
As for the asynchronism, it is still in favour of the CO bond.

We have recalculated the total energy of the supersystem for different values of R and τ for the three types of CI mentioned in the
above sections. They include a (3 x 3), (15 x 15), and (55 x 55) CI for the
mono- and diexcitations between one, two, and three pairs of HOMO's
and LUMO's, respectively. We have retained the other parameters optimized in STO-3G. The results of the calculations with configuration
interaction are presented in table 11. As could be expected, the geometry of the activated complex is modified by CI. However, whatever be
the CI used, the approach C is always less favorable. The variation
of the total energy in function of the angle τ for R = 2.25 Å indicates that the structure of the activated complex is even closer to the
approach P ($\tau \simeq 70°$) than C.

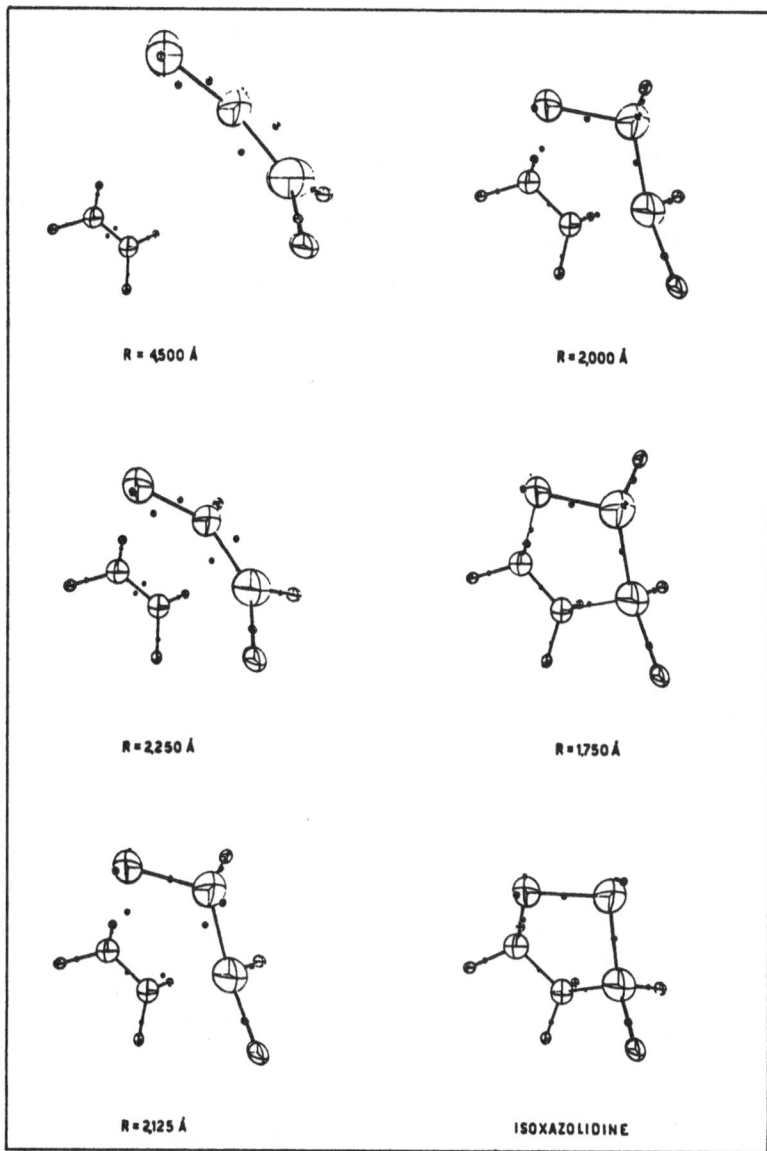

Figure 15. Evolution of charge centroids along the reaction path for ni-
     trone to.ethylene.

There is good reason then to believe that the SCF results are qualitatively correct. In particular, the type of approach of the reactants
and the geometry of the transition state can be considered significant.
The energetic characteristics of the reaction vary, of course, from one
calculation to another but one will notice that with a limited CI the
activation energy is of the order of 30 kcal/mole and the heat of reaction around - 80 kcal/mole.

Table 11. Total energy of the supersystem nitrone-ethylene around the
         transition state.

Method	$R = \infty$	$\tau = 90°(P)$	R = 2.25 Å $\tau = 39°$	$\tau = 0°(C)$	Isoxazolidine
SCF	-243.638862	-243.597075	-243.592622	-243.544847	-243.858363
(3 x 3) CI	-243.691548	-243.644574	-243.647289	-243.612701	-243.858857
(15 x 15) CI	-243.736081	-243.677838	-243.673252	-243.634425	-243.859948
(55 x 55) CI	-243.736967	-243.678854	-243.683098	-243.656852	-243.862930

Overall, despite their approximated and qualitative character, the minimal basis set SCF results obtained in this work permit one to pick
out the principal characteristics of the concerted cycloaddition of nitrone with ethylene :

   - until the transition state (R = 2.25 Å) the approach of the reac-
     tants is nearly in two parallel planes and their structure -
     electronic as well as geometric – is very little changed ;
   - starting at the transition state one finds an asynchronism in
     the formation of the new bonds : $T_{CO} = 8.2\%$ and $T_{CC} = 3.7\%$;
   - the charge transfer towards ethylene is maximum at this point ;
   - the activation barrier equals 26 kcal/mole in STO-3G , 35 kcal/
     mole in 4-31G and around 34 kcal/mole with a (55 x 55) CI (with
     STO-3G) ;
   - beyond the transition state the major electronic reorganization
     begins : the "π electrons" of the $N_4 \cdots O_5$ and $C_1 \cdots C_2$ bonds
     migrate into σ loges of the $C_1 \cdots O_5$ and $C_2 \cdots C_3$ bonds in for-
     mation ; one of the centroids of the $C_3 \cdots N_4$ bond is displaced
     towards the pseudoequatorial position which the centroid of the
     nitrogen lone pair will occupy in the product.

3.1.3. Discussion.

    Our theoretical study of the 1,3-dipolar cycloadditions should be
considered as only preliminary due to the imprecision of the calculations tied in with the rather large limitation of the basis sets, and
due to the small size of the configuration interactions employed. It
then follows that only the relative values of the energies obtained
within one given framework can have a meaning. Likewise, one can retain
only the structural characteristics of the mode in which the reactants
approach each other. As for the results of the different levels of configuration interactions, they show that the values calculated with the
STO-3G basis set in the RHF-SCF-LCAO-MO method are at least qualitatively correct.
    We have at our disposal, then, a large body of quite coherent re-
sults on the 1,3-dipolar cycloadditions which permit us now to draw
several general conclusions.

With the view to facilitating the discussion, we have assembled, in
table 12 the principal characteristics of the reactions calculated here.
     According to the mode of approach for the reactants one can dis-
tinguish two classes of dipoles, which correspond to those proposed
by Huisgen [11] :

- the dipoles of the first type, or the "propargyl-allenyl type"
  with 22 electrons, are linear (not including the hydrogen atoms)
  and possess either 6 or 8 π and $\bar{\pi}$ electrons where four are delocalized on three centers. Some examples of this type of dipole are diazomethane, nitrous oxide and fulminic acid ;
- the dipoles of the second type, or "allyl type" with 24 electrons,
  possess only 4 π electrons which are delocalized over 3 centers
  and are bent. In their cycloadditions to ethylene the reactants
  approach in two different planes, parallel or not. As such, the
  5 heavy atoms are not in the same plane at the transition state.
  Some examples of this type of dipole are ozone, carbonyl ylide,
  and nitrone.

In general, the geometry of the activated complex is rather similar to
that of the reactants. This resemblance can be qualitatively estimated
by the value of R at the transition state for each dipole. This structure of the activated complex is seen not to be appreciably affected
by configuration interaction.
     We recall the geometries of the activated complexes in figure 16
for the different 1,3-dipolar cycloadditions that we calculated in the
STO-3G basis set. Thus, with carbonyl ylide and nitrone $\tau = 90°$ and with
ozone $\tau = 63°$. In the case of diazoalkanes, nitrous oxide, and fulminic
acid $\tau = 0°$.
     The absence of hydrogen atoms in the molecule of ozone might be
the cause of the smaller value of the angle τ in the activated complex
where this dipole intervenes.
     We certainly do not have at our disposal as yet a sufficient number
of entirely reliable results in order to give a satisfactory interpretation
of the variations in the envelope angle at the transition state ($\tau^{\mp}$) in function of the electronic and geometric structure of the dipoles.

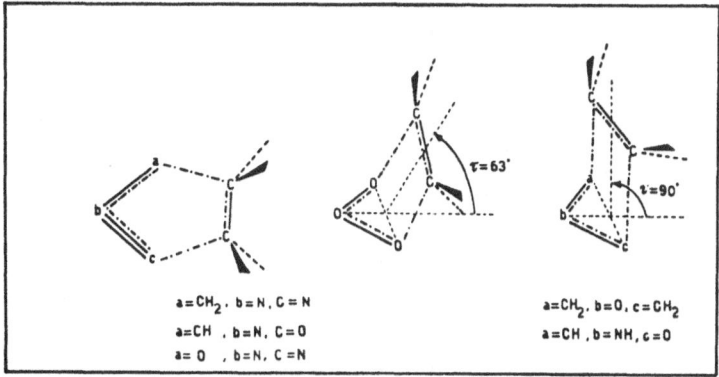

Figure 16. Structures of the activated complexes in 1,3-dipolar cyclo-
           additions.

Table 12. Characteristics of the 1,3-dipolar cycloadditions studied in this work.

Reaction	$R^{\#}$(Å) STO-3G	$\tau^{\#}$(°) STO-3G	$\Delta E^{\#}$(kcal/mole)				ΔE_o(kcal/mole)				ΔH_o^{o}(kcal/mole)	$t^{\#}(e^-)$ STO-3G	Mode	$A^{\#}$(Z) STO-3G
			STO-3G	7s-3p	4-31G	(55×55) CI	STO-3G	7s-3p	4-31G	(55×55) CI				
$C_2H_4 + CH_2N_2$	2.25	0	21.2	19.4	26.2	–	–111.6	– 82.5	– 29.2	–	–27	0.06	C	10.1(C-C)
$C_2H_4 + CHNO$ *	2.30	0	15.5	18.2	28.8	–	–132.8	– 83.7	– 50.1	–	–	0.02	C	30.3(C-O)
$C_2H_4 + N_2O$	2.15	0	21.4	23.3	29.2	31.3	– 93.8	– 44.6	– 14.5	– 37.2	–	–0.01	C	28.3(C-O)
$C_2H_4 + CH_2OCH_2$	2.35	90	6.0	1.0	10.3	–	–181.2	–152.0	–114.6	–	–	0.09	P	0.0
$CH_2O + CH_2OCH_2$	2.15	90	3.9	10.9	5.7	17.7	–171.1	–116.3	–106.2	–105.2	–	0.16	P	1.3(C-O)
$C_2H_4 + O_3$	2.30	63	1.3	5.2	1.7	–	–136.8	– 96.9	– 92.5	–	–49	–0.02	P	0.0
$C_2H_4 + CH_2NHO$	2.25	90	26.3	24.6	35.5	34.0	–137.7	– 77.6	– 39.6	– 80.0	–55	0.10	P	37.8(C-O)

* D.Poppinger - Personal communication

One will note, however, that in the P approach of carbonyl ylide to
ethylene the electrostatic potential of the supermolecule presents positive values in the regions of the formation of new bonds, whereas,
in the C approach, it presents negative values for these regions. This
might be at this point only a simple coincidence, but one finds the
same result in the cycloaddition of nitrone to ethylene. This is illustrated in figure 17. It thus seems that the approach of the reactants
is carried out in a manner that creates zones of low potential for the
electrons in the intermolecular regions.

Furthermore, we find that no apparent correlation exists between
the activation barriers and the modes of approach of the reactants.

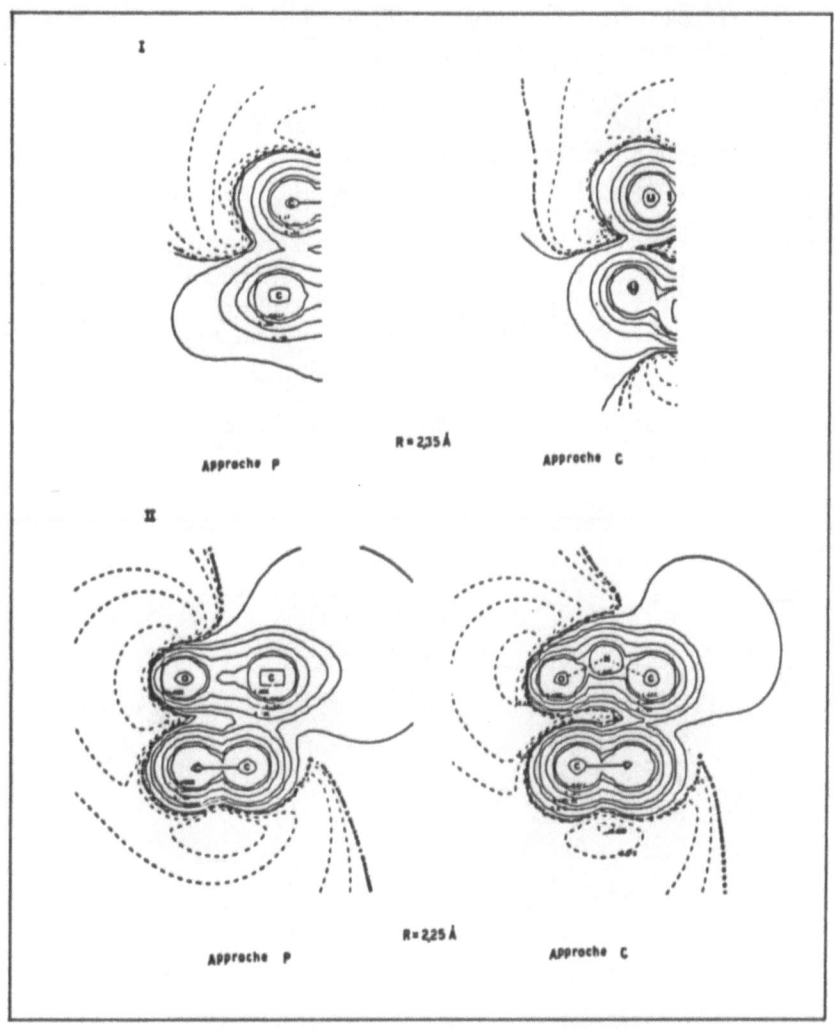

Figure 17. Electrostatic potential for the supersystems of ethylene
plus carbonyl ylide (I) and nitrone plus ethylene (II).

We note at this point that the results obtained in the 4-31G basis set
are subject to caution since the calculations have been carried out on
the geometries optimized in STO-3G. The following points are to be made
concerning the electronic reorganization :

- the direction of the migrations of the bonds seems to be determined by the charges of the terminal atoms. Actually, with un-
  symmetrical dipoles, the movements of the centroids start from
  that end of the dipole with the most negative heavy atom, and
  it is also the intersystem bond relative to this atom which is
  the most precocious. The figure 18 illustrates the findings ;
- the transfer of charge is rather small is all cases and is most
  often in the direction from the dipole towards the dipolarophile.
  It allows one to predict the influence of substituents on the
  reactivity of the two partners.

In total, at this point of our study of 1,3-dipolar cycloadditions we
have shown that the approach of the reactants, and thus the structure
of the activated complex, depends on the electronic and geometric structure of the dipole. This thus gives solid theoretical support to the
classification proposed by Huisgen[11].

Our work equally leads to a detailed description of bond reorganizations which is better founded than those which relie on resonance theory.

We have not yet, however, found any inherent reason for the difference in the behavior of the first and second types of dipoles.

A refinement remains for our results in using more exact wave-func-
tions and the generalization of our results remains by exploring the portions of the potential energy hypersurfaces corresponding to non-concerted approaches.

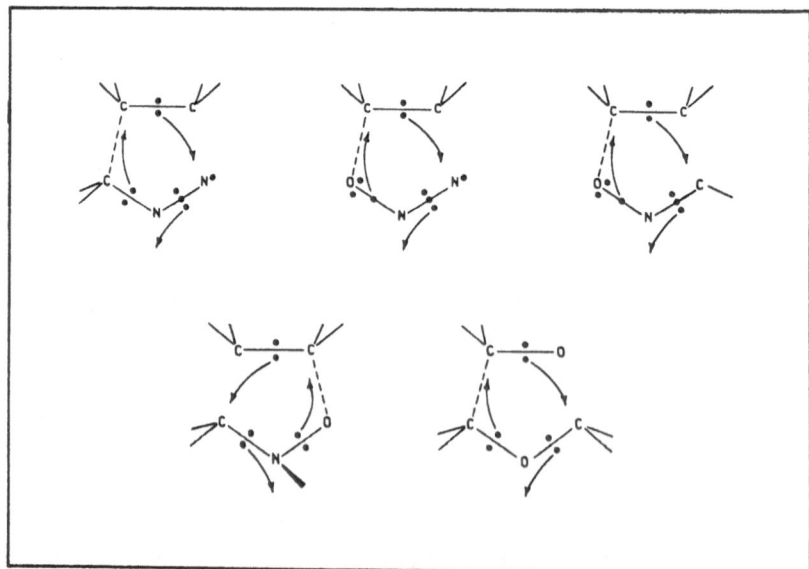

Figure 18. Electronic reorganization in the studied 1,3-dipolar cyclo-
           additions.

3.2. Ring-chain isomerizations.

3.2.1. The azido-tetrazole isomerization[30]

   We have studied the mechanism of the isomerization of azidoazomethine to 1H-tetrazole using the same theoretical approach as the one
used for the 1,3-dipolar cycloadditions. The azidoazomethine-tetrazole
isomerization is a member of an important family of ring-chain isomerizations as given below.

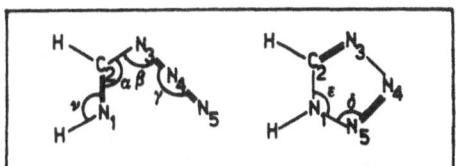

Y = N	X = N	: the present reaction
Y = N	X = CH	: see 3.2.2.[31]
Y = N	X = $\overset{+}{N}$H	: see 3.2.3.
Y = CH	X = N[32]	
Y = CH	X = CH[33]	

The geometric parameters used in the optimization procedure are defined in figure 19. The distance between the atoms N_1 and N_5 was chosen
as the reaction coordinate for constructing the reaction path.

Figure 19. Geometric parameters.

   The principal results of this study can be summed up as follows. The
cyclization process has essentially three steps that can be described
in geometric, electronic, and energetic terms. The first step, which
continues up to the transition state, is characterized by a diminishing
of angle γ along with the lengthening of bonds N_3N_4 and N_4N_5. During
this step one of the N_4N_5 triple bond centroids moves to form a lone
pair on N_4. The concomitant rise in energy is relatively small (12.3
kcal/mole with STO-3G and 5.0 kcal/mole with the 7s-3p basis set). This
can be taken for the activation energy for the cyclization reaction.
   The second step corresponds to diminishing of the angle α to the
value found in tetrazole. The angles γ and ν vary simultaneously but
the only bond length which undergoes an appreciable modification during
this step is N_4N_5. Parallel to these geometric changes, one of the N_3N_4
centroids starts to migrate towards bond C_2N_3, while the system begins
to stabilize appreciably.

It is during the third step that the most important modifications manifest themselves in a geometric, electronic, and energetic point of view. All the structural parameters take up the values corresponding to tetrazole ; the σ bond between atoms N_1 and N_5 is formed at the expense of the lone pair on N_1 while the π electrons of the N_3N_4 finally form the double bond on C_2N_3. In the last place, two π electrons of the C_2N_1 bond become the electron pair on N_1 of tetrazole. All these results are summarized in figure 20. A keynote of the whole process is the role of the lone pair on N_1 which permits the formation of the new sigma bond without having to turn the π_{12} system.

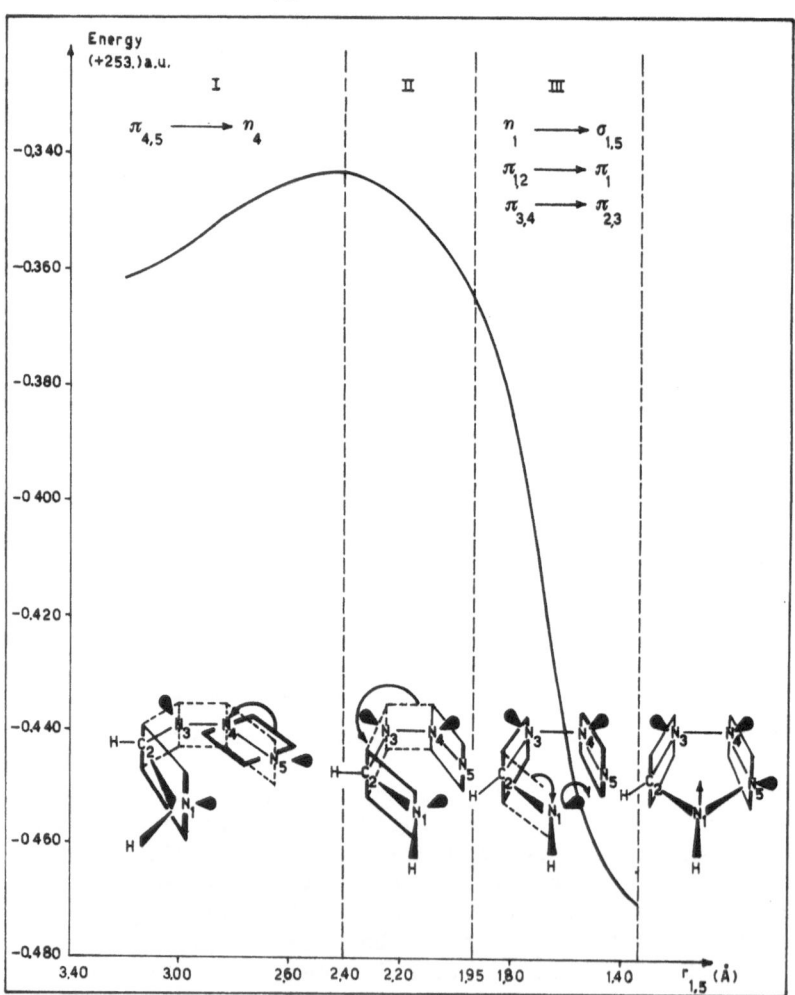

Figure 20. Characteristics of the reaction path for the azido-tetrazole
isomerization.

In conclusion, one can say qualitatively that the activation energy for the cyclization process is due essentially to the bending of the angle γ while that for the reverse process is due to the breaking of the bond between N_1 and N_5 and to the loss of the delocalization energy in tetrazole. Finally, let us note that no precise experimental data have ever been obtained for the isomerization of azidoazomethine to 1H-tetrazole ; indeed, tetrazole is the sole form observed, even in the vapor phase at 93°C[34]. However, an experimental value of 17.8 kcal/mole was found for the activation energy of the cyclization of guanylazide into amino-tetrazole[35]. In any case, accounting for the effects of substituents and solvents, temperature and ZPE corrections, the theoretical values for the energetics, as given in figure 20, can hardly be compared with any experimental results.

3.2.2. The vinyl azide-v-triazole isomerization[31].

It was of interest to compare the cyclization mechanism of vinyl azide to v-triazole to that for the azido-tetrazole isomerization des-cribed in the above section.

The geometric parameters for the optimization of the vinyl azide closure are illustred in figure 21.

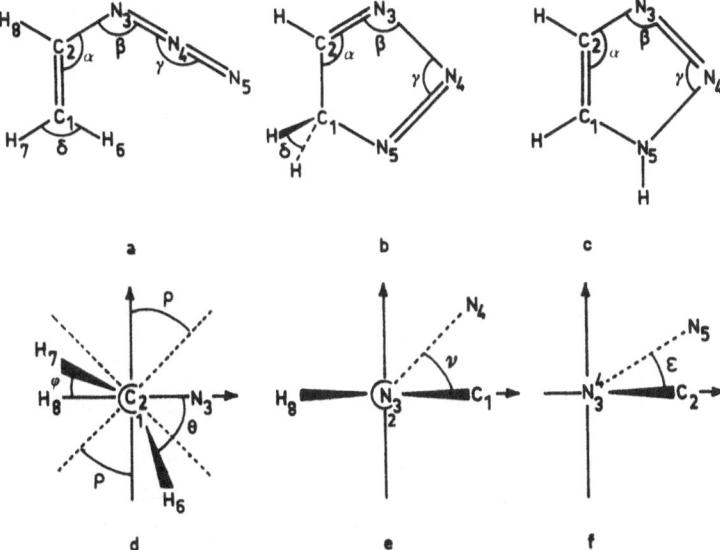

Figure 21. Geometric parameters for the optimization along the vinyl azide closure reaction path.

The dihedral angles considered are :

$$\begin{aligned}
\theta &= H_6 C_1 C_2 N_3 \\
\phi &= H_7 C_1 C_2 H_8 \\
\nu &= C_1 C_2 N_3 N_4 \\
\varepsilon &= C_2 N_3 N_4 N_5
\end{aligned}$$, as shown in figure 21 (d → f).

The only constraint upon the angles was that atoms C_1, C_2, N_3 and H_8 remain planar and that the line C_2H_8 bisect angle γ. The only bond dis-tances not varied were the CH distances (1.08 Å).

The reaction coordinate was chosen as the C_1N_5 distance for the sole reason that it is across this space that a new bond is formed.

An analysis of the calculated reaction path demonstrates the existence of three zones in the cyclization process.

The first can be called the azide zone since the supermolecule resembles the reactant, albeit an azide bent at N_4. The second is a transition zone in which the CH_2 group begins to rotate. The measure of this rotation, ρ (figure 21 d), is 25° at the transition state and 44° at the end of the second zone. In the third, which can be called the triazole zone, the system strongly resembles the product v-triazole.

This persistance of either π system and a narrow region in which the CH_2 group rotates has already been shown in the ab initio calculations on the electrocyclic reactions of 1,3-butadiene[36] and of acrolein[37].

The reorganization of the chemical bonds along the reaction path can be described by the displacements of the centroids of charge for localized molecular orbitals. One finds that the azide zone is a process in which a lone pair is formed on N_4 while the π system for the molecule scarsely changes. In the transition zone, most of the major electronic reorganization takes place. Finally, in the third zone the bond between C_1 and N_5 is formed.

The activation energy and the heat of reaction for the cyclization of vinyl azide to v-triazole using the minimal STO-3G basis set are respectively 32.9 kcal/mole and - 56.1 kcal/mole. The principle characteristics of this reaction are summarized in figure 22. The activation energy for the closure of vinyl azide is due to the bending of the angle γ and to the rupture of the π system by turning of the terminal CH_2 group, and that of the reverse process is essentially due to the breaking of the bond between C_1 and N_5. The minimal basis set activation energy of 32.9 kcal/mole for the closure process agrees rather well with the experimental activation energies for the decomposition of several substituted vinyl azides (26-30 kcal/mole)[32]. This result is fortuitous as the extended basis set 4-31G indicates a value around 40 kcal/mole.

We have tested the influence of limited configuration interactions on the energy hypersurface. The configurations were constructed from the mono- and diexcitations from the highest occupied molecular orbital (HOMO) to the lowest unoccupied (LUMO) (3 configurations), from the 2 HOMO's to the 2 LUMO's (15 configurations), and from the 3 HOMO's ti the 3 LUMO's (55 configurations). We found that in all cases the geometries of the activated complex and the activation energy were only slightly lowered. It thus seems that the SCF procedure alone is sufficient to give qualitative features of the vinyl azide cyclization reaction.

3.2.3. The cyclization of protonated azido-azomethine.

The geometries found for the vinyl azide cyclization were used as starting points to study the closing of protonated azido-azomethine to protonated tetrazole.

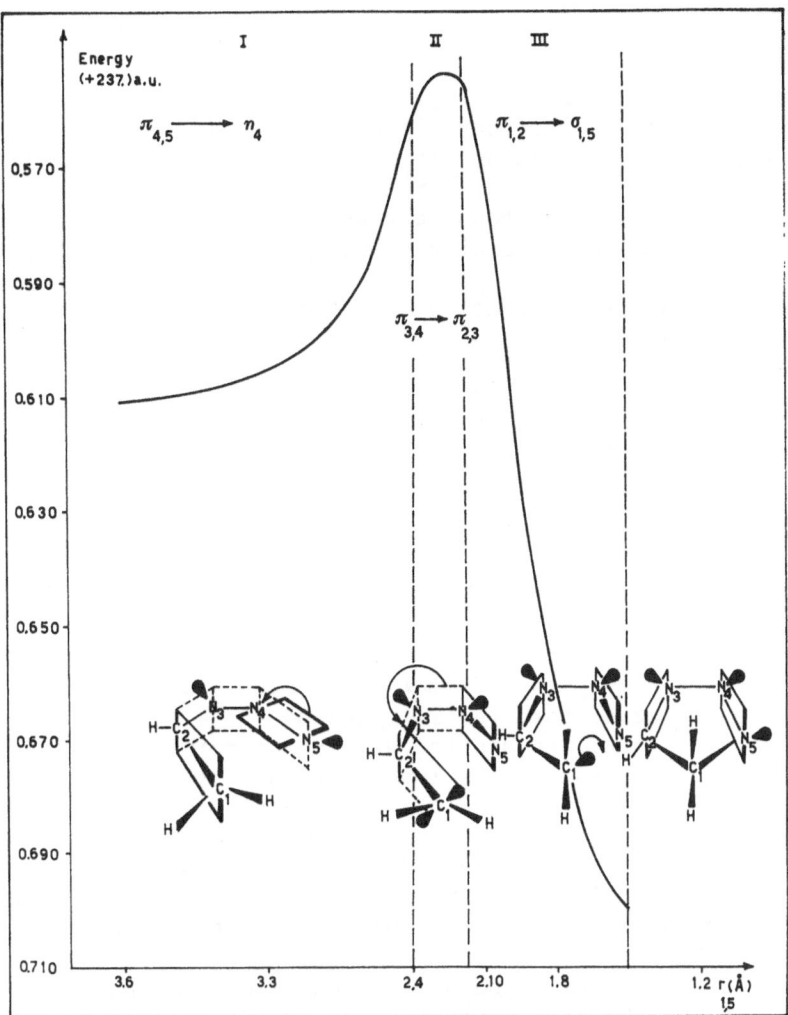

Figure 22. Characteristics of the reaction path for the vinyl azide-v-triazole isomerization.

In the previous study of the azido-tetrazole isomerization, it was found that the reaction proceeds by an attack of the NH lone pair on the other terminal nitrogen. Protonation of this lone pair by acidifying the media would lead to another type of reaction, such as an electrocyclic reaction in which the π system must be disrupted by turning of the terminal NH_2^+ group. This might then give a higher activation energy than the planar reaction by the non-protonated species.

It was found, after the optimization procedure, that the geometries for protonated azido-azomethine were the same, within $2°$ and $0.02\,\text{Å}$, as their vinyl azide counterparts. The activation energy is 42.9 kcal/mole using the STO-3G basis set.

The cyclizations of protonated azido-azomethine and vinyl azide present enough structural and electronic similarities to be considered as the same type of reaction.

3.2.4. Discussion.

The cyclizations of vinyl azide and protonated azido-azomethine can be classed as electrocyclic reactions. However, they present electronic properties which permit a comparison with the cyclization of neutral azido-azomethine. The comparison is presented below by means of the principal types of bond reorganizations based on the displacements of localized molecular orbital centroids of charge during the reaction.

Azido tetrazole isomerization 8 electrons involved		Vinyl azide triazole isomerization 6 electrons involved	
Zone I	$\pi_{4,5} \longrightarrow n_4$	Zone I	$\pi_{4,5} \longrightarrow n_4$
Zone II	—	Zone II	$\pi_{3,4} \longrightarrow \pi_{2,3}$
Zone III	$\pi_{3,4} \longrightarrow \pi_{2,3}$	Zone III	$\pi_{1,2}$
	$\pi_{1,2} \Longrightarrow \pi_1$		$\searrow$
	$n_1 \longrightarrow \sigma_{1,5}$		$\sigma_{1,5}$

In the two cases the formation of a lone pair is observed on N_4 at the expense of the orthogonal $\pi_{4,5}$ bond, and this reorganization takes place well before the transition state (in the azide zone). The activated complex is very similar (from the electronic and geometric points of view) to the azide form in all cases.

For its part, the neutral azido-azomethine cyclization procedes via the displacement of the lone pair on atom N_1 towards atom N_5. This attack is the driving force for the reaction and the reorganization of the π system is a simple consequence of this first electronic movement. Protonated azido-azomethine and vinyl azide do not possess this lone pair and thus the π system must participate directly in the formation of the σ bond between atoms 1 and 5. It is this rupture of the π system which is responsible for the higher activation energy in the cyclizations of vinyl azide and protonated azido-azomethine.

Finally, the mechanisms for the cyclization reactions studied here can be described by the following formulae :

3.3. Symmetric cycloadditions.

A cycloaddition between partners which contain each an axis of at least C_2 symmetry will be called symmetric even though the activated

complex might not contain a C_2 axis of symmetry. The two cases studied
here will be the concerted Diels–Alder reaction between ethylene and
cis–butadiene and the concerted dimerization of ethylene. The addition
of symmetric dipoles to ethylene is also a member of this group but has
been studied in the section on 1,3–dipolar cycloadditions.

3.3.1. A Diels–Alder reaction[38-39].

We have studied the concerted addition of ethylene to cis–butadiene which is an archetype of the Diels–Alder reaction between olefins
and dienes. The product, cyclohexene is most stable in its half–chair
conformation, and so care was taken to introduce a parameter in the optimization of the supermolecule which would account for a conversion in
the transition state between a boat to half–chair form.

The numbering of the atoms as well as some of the angles that were
optimized are presented in figure 23. All parameters were optimized for
several chosen values of R_{16} and R_{45}. These two parameters, which are
equal as is required for a "concerted" reaction, were chosen as the
reaction coordinate. Besides the angles shown in figure 23 several other
angles were optimized :

δ : retreat of H_{14} and H_{16} ;
ε : retreat of H_{13} and H_{15} ;
μ : lowering of H_{10} ;
ν : lowering of H_{11} .

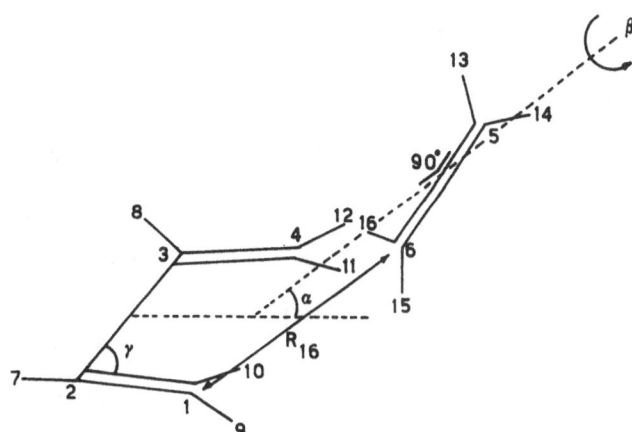

Figure 23. Geometric parameters used in the calculations.

There exists an interdependence of the parameters μ and ν with the bond
lengths[39]. The lower H_{10} and H_{11}, the closer are the bond lengths to
those in the product. Care has to be taken to optimize these parameters
iteratively[39].

The method of calculation employed for the optimizations was the
STO–3G minimal basis set.

The transition state was also optimized for μ, ν, and the bond lengths using four types of CI. The most extended of these was a (55 x 55) CI involving all mono- and diexcitations from the three highest occupied molecular orbitals (HOMO) to the three lowest unoccupied (LUMO). A (15 x 15) CI involving the two HOMO's and the two LUMO's was used, as well as two different (3 x 3) CI's. The first involves excitations from the occupied MO which is a combination of the butadiene π_2 MO with increasing amounts of the ethylene π^* MO to the corresponding unoccupied MO (butadiene π_3 with increasing ethylene π). This is the (3 x 3) CI involving the MO's developed along the reaction path from the HOMO-LUMO combination at large separation between ethylene and butadiene (R_{16}) and was used by Salem and coworkers[40]. The other (3 x 3) CI involves the next HOMO with the next LUMO described at large R_{16}. The next HOMO at large R_{16} is essentially the ethylene π MO but with increasing amounts of butadiene π_3 and π_1; and the next LUMO is the ethylene π^* with increasing butadiene π_2 and π_4. It is necessary to specify the origin of the HOMO and next HOMO since the two MO's cross near the transition state. The geometric parameters for points optimized in these methods are given in table 13.

Figure 24 is a composite of the energies of the two HOMO's and the two LUMO's along with the relative energies of the supermolecule along the reaction coordinate, R_{16}. The zero point for the relative energies using the different CI methods is the total energy of the supermolecule at $R_{16} = 5.0$ Å. The solid lines for the MO energies are for the supermolecule whose geometries were optimized with the single determinant SCF method. When the supermolecule was optimized using the first described (3 x 3) CI, different geometries than those for the single determinant SCF were found for the supermolecule in the region of the transition state (table 13). The energies of all the MO's remained fairly similar for a given R_{16} with the two methods, except, however, for the MO's derived from the butadiene π_2 and π_3 MO's (which were used for this (3 x 3)CI). One observes that, using this CI in the optimization, geometries were found where the energy gap is narrowed between the two molecular orbitals with which the CI was constructed.

When the supermolecule is optimized using a (3 x 3) CI which uses the ethylene π MO's, the geometries found are the same as those found in the single determinant SCF method.

The dependence of the butadiene CI results on one geometric form and the independence of the ethylene CI results may be explained in the following way. The π_2 orbital has a node between C_2 and C_3, and any shortening of the C_2C_3 bond length towards product will destabilize this orbital. Likewise, the shortening will stabilize the π_3^* orbital since there is maximum overlap between C_2 and C_3 for this orbital. As the ethylene π orbital correlates with the cyclohexene π orbital, the molecular orbital coefficients in the transition state show a bonding between C_2 and C_3 nearly equal to that between the C_5 and C_6 of ethylene. If one chooses a geometry close to cyclohexene, the stabilization by shortening C_2-C_3 is counterbalanced by the destabilization by lengthening C_5-C_6. A similar counterbalance can be shown for this molecular orbital if a reactant-like geometry is chosen for the transition state.

Table 13. Geometric parameters for points along several reaction paths : a, SCF ; b, (55 x 55) CI ; c, (15 x 15) CI ; d, (3 x 3) CI (butadiene) ; e, (3 x 3) CI (ethylene).

State	$R_{16}=R_{45}$(Å)	R_{12}(Å)	R_{23}(Å)	R_{56}(Å)	α(°)	β(°)	γ(°)	ϵ(°)	δ(°)	μ(°)	ν(°)	E_T(a.u.)
I[a]	5.000	1.314	1.497	1.310	72	0	126	0	0	0	0	-230.088419
b	5.000	1.314	1.497	1.310	72	0	126	0	0	0	0	-230.193155
c	5.000	1.314	1.497	1.310	72	0	126	0	0	0	0	-230.158991
d	5.000	1.314	1.497	1.310	72	0	126	0	0	0	0	-230.104697
e	5.000	1.314	1.497	1.310	72	0	126	0	0	0	0	-230.130080
II[a]	3.167	1.314	1.497	1.310	72	0	126	0	0	0	0	-230.087223
III[a]	2.867	1.314	1.497	1.310	72	0	126	0	0	0	0	-230.077537
IV[a]	2.567	1.314	1.497	1.310	72	0	123	0	0	0	0	-230.058055
V[a]	2.367	1.330	1.465	1.330	70	0	123	5	5	15	15	-230.038375
VI[a]	2.300	1.350	1.455	1.350	66	5	120	12	8	22	25	-230.024376
d	2.300	1.375	1.410	1.380	62	5	120	17	13	28	30	-230.043696
VII[a]	2.240	1.375	1.410	1.380	62	5	120	17	13	30	32	-230.024174
b	2.240	1.400	1.360	1.410	62	5	120	17	13	30	32	-230.124169
c	2.240	1.375	1.410	1.380	62	5	120	17	13	30	32	-230.079256
d	2.240	1.400	1.360	1.400	62	5	120	17	13	35	37	-230.043645
e	2.240	1.375	1.410	1.380	62	5	120	17	13	30	32	-230.040931
VIII[a]	2.200	1.375	1.410	1.380	62	5	120	17	13	35	37	-230.026651
IX[a]	2.167	1.400	1.360	1.410	62	5	120	17	.13	35	37	-230.031761
X[a]	1.504	1.504	1.325	1.542	0	15	122	0	120	105	135	-230.255489

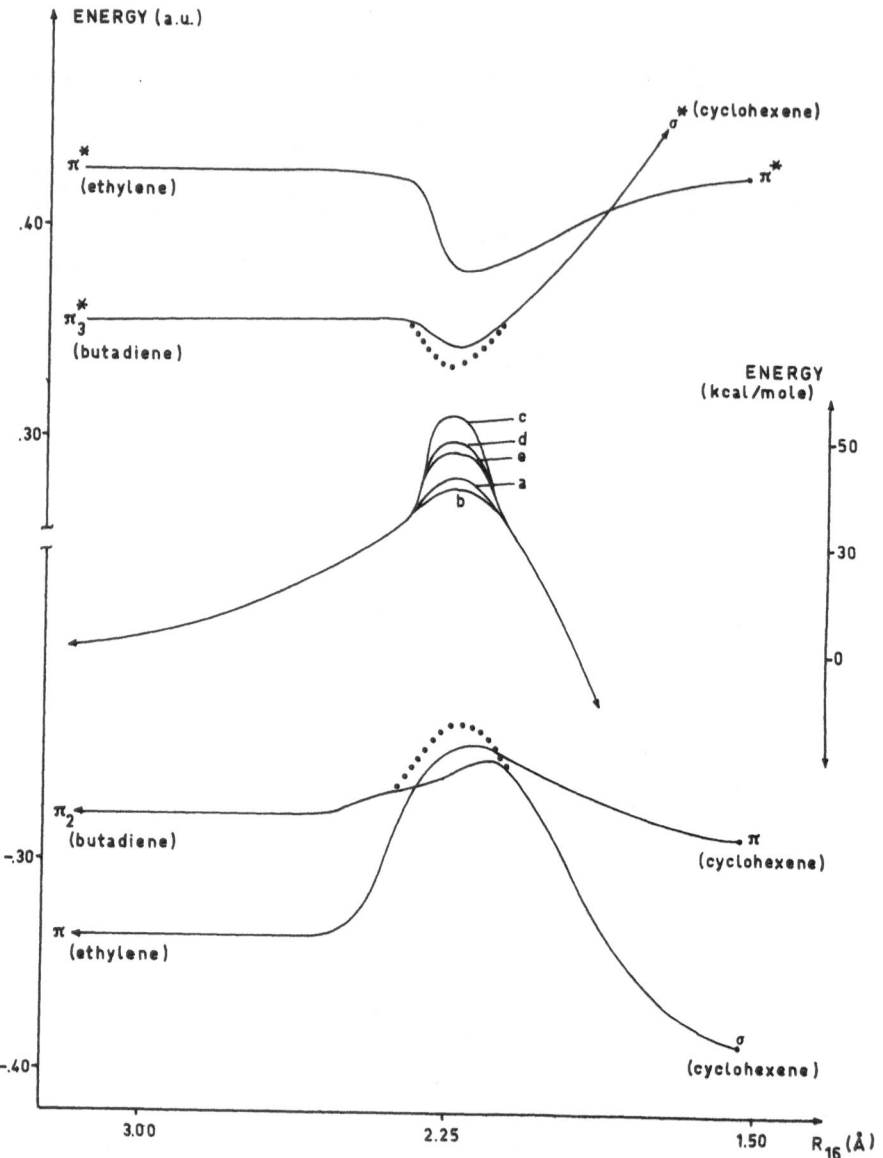

Figure 24. Superposition of two graphs along the reaction coordinate, R_{16} (Å). In the middle (ordinate at right) is the relative energy of the optimized supermolecule with the value at $R_{16} = 5.0$ taken as 0.0 kcal/mole (a, SCF ; b, (3×3) CI (π but.) ; c, (3×3) CI (π eth.) ; d, (15×15) CI ; e, (55×55) CI). The energies of the two HOMO's and two LUMO's (ordinate at left) are also given (see text).

Therefore, a (3×3) CI using this orbital does not favor the geometry of either product or reactants but rather an intermediate geometry.

Choosing one or the other (3×3) CI might also be questioned when calculating activation energies. Since an activation energy is the difference between the total energies for the transition state and a point far out on the reaction coordinate, one is taking the difference between two different types of (3 x 3) CI's. At large values of R_{16}, the excitations are from an essentially pure butadiene π_2 to its π_3^* MO ; at the transition state the MO from which the excitations come is a mixture of the butadiene π_2 MO and the ethylene π^* MO. The same argument can be used for the other (3 x 3) CI where at large R_{16} the excitations are from the ethylene π MO and at the transition state it is from a MO which is a mixture of the ethylene π MO and the butadiene π_1 and π_3^* MO's. It might seem more reasonable to use a (55 x 55) CI where the MO's used in the transition state are all used at large R_{16} as well.

This then means that the activation energy is higher using a large CI than it is with the single determinant SCF method (47.4 (55 x 55) , 50.0 (15 x 15) , 55.9 (3 x 3)ethylene , 38.3 (3 x 3)butadiene , 40.3 kcal/ mole (SCF)). Basilevsky and coworkers[41] also found a raising of the barrier for this reaction when electron correlation terms were included in their version of a PPP semi-empirical method which is modified to account for exchange repulsion.

Results based solely on the SCF method are quite sufficient to describe this reaction since this configuration contributes more than 90 % to the 55 configurational expansion. As such the charge transfer results based on a Mulliken population analysis can be significant. It was found that a transfer of 0.021 electron towards ethylene is a maximum which developes just before the transition state, at $R_{16} = 2.30$ Å.

The geometric parameters show that the transition state is intermediate between reactant – and product – like geometries. Indeed, the transition between the boat and half-chair forms has already started before that point. The centroids of charge (figure 25), however, show that a reactant – like positioning is obtained even after the transition state (point VIII, in figure 25). A product – like positioning is only attained at point IX ($R_{12} = 2.167$ Å). It is to be remarked that the centroids give only reactant – and product – like positionings and an intermediate positioning was never seen.

In conclusion, for this reaction, we note two ways in which the calculation of transition states can be influenced. The first, interdependence of variables, is procedural and can lead to traps on the hypersurface for the unwary. The second, the choice of limited CI's, concerns the method used ; one must be wary of whether the method will favorize one geometric form or another.

3.3.2. The ethylene dimerization reaction.

Although the thermal dimerization of ethylene has never been achieved, the thermal dissociation of cyclobutane has been shown to be a first order reaction with an activation barrier of 62.5 kcal/mole[42]. Furthermore, olefins have been found to add to highly activated ethylenes such as tetracyanoethylene (TCNE) in a wide range of organic solvents[43]. This latter class of olefinic dimerizations has been found to proceed with retention of configuration ; that is, a cis or trans olefin will add to TCNE to form a cis or trans cyclobutane, respectively, to a much larger degree than the trans or cis isomer.

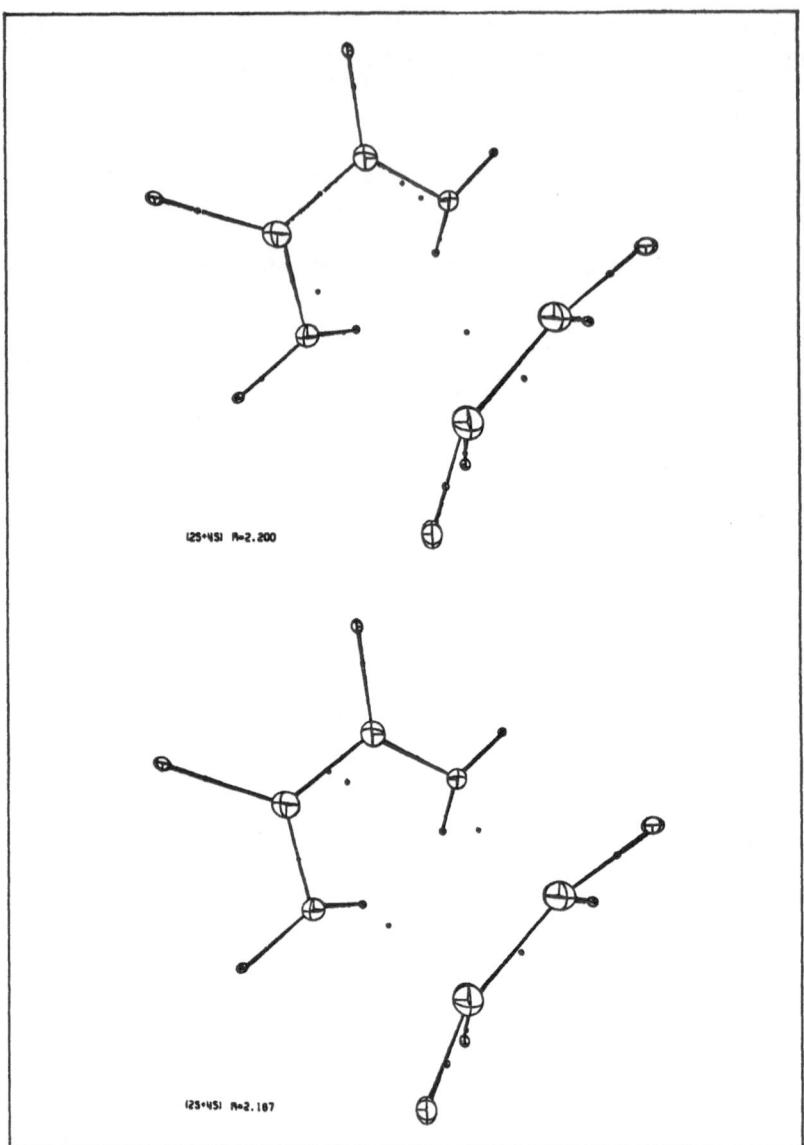

Figure 25. Evolution of the centroids along the reaction path. The reactant‑like position is found for point VIII (R = 2.20 Å) and the product‑like for point IX (R = 2.167 Å) of table 13.

Thus, the thermally forbidden[44], 2 π suprafacial + 2 π suprafacial (2 π_s + 2 π_s) cycloaddition is suspected to take place above all in this type of ionic olefinic dimerization. On the other hand, the minor cyclo‑inverted isomer products are expected to be formed via a two‑step mechanism where a diradical or zwitterionic intermediate is involved. This two‑step mechanism is expected for lightly substituted ethylenes in solvents or for all olefins in the gas phase.

In an ab initio, STO-3G, (15 x 15) CI study of the two - step mechanism, G. Segal[8] found the activation barrier for the dimerization of
ethylene to be 42.4 kcal/mole. It would seem that this mechanism is
highly favored in the gas phase since Salem and Wright[45] found an activation barrier near 128 kcal/mole for the (2 π_s + 2 π_s) cycloaddition
reaction employing the STO-3G, (2 x 2) CI method. The use of a (15 x 15)
CI might, however, lower this barrier.

There exists another concerted mode of approach, the 2 π suprafacial + 2 π antarafacial (2 π_s + 2 π_a), in which the antarafacial
partner undergoes inversion of its substituents upon addition. This approach is thermally allowed since there exists no symmetry elements
which would permit an avoided crossing between two molecular orbital
configurations[46]. However, there is expected to be a large amount of
steric repulsion with an antarafacial approach where an ethylene is added edgewise. Two classes of molecules are expected to undergo a (2 π_s
+ 2 π_a) addition, however. The first contains olefins with highly
strained double bonds such as 1 [47] and the second contains the cumulenes such as ketene 2 [48]. Ketenes present a central carbon atom at
which there are two orthogonal π systems. This is expected to ease the
antarafacial approach of ketene since there is no need to twist two
CH_2 groups apart in order to cycloadd, as in ethylene.

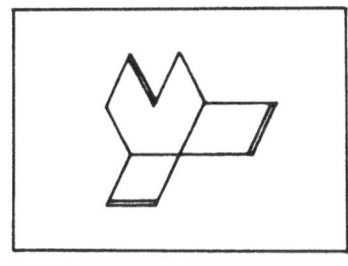

$$H_2C = C = O$$

1   2

A potential energy hypersurface was constructed to include the two concerted paths in the STO-3G, (15 x 15) CI method in order to situate the
(2 π_s + 2 π_a) path vis-a-vis the (2 π_s + 2 π_s) (which was only explored
in the region of the transition state by Salem and Wright[45]) and also
the concerted barriers vis-a-vis the two step barrier.

The major constraint in the geometry optimization for the concerted (2 π_s + 2 π_a) ethylene dimerization pathway was that the two CC
bonds being formed would always be of equal length. The reaction coordinate was chosen to be this distance between the terminal carbon atoms
solely because it is across these two spaces that new bonds are formed.
Another perspective is that for any given point on the reaction coordinate the optimization procedure would start from the suprafacial-anta-
rafacial (SA) position of the supermolecule. In this way one might show
the existence of a form near the idealized SA configuration by demons-
trating the existence of a barrier between this form and the suprafa-
cial-suprafacial form.

The full set of geometric parameters was optimized iteratively in case an interdependence of parameters exists. The numbering of the carbon atoms and the angles that were optimized are shown in figure 26. The order is as follows :

α : rotation about a central axis ($\alpha = 0$ for $2\,\pi_s + 2\,\pi_s$) ;
β : retreat of a hydrogen (pyramidalization = $0.5\,\beta$) ;
γ : rotation of both hydrogens ;
δ : angle HCH .

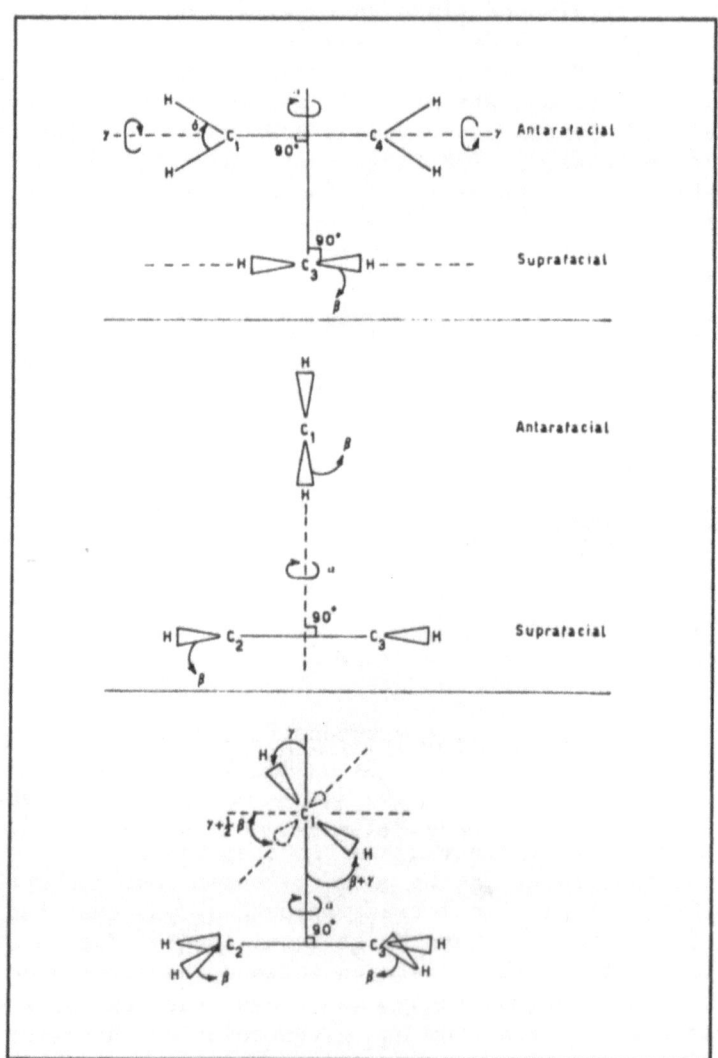

Figure 26. The geometric parameters in the ethylene dimerization reaction.

The twisting apart of the CH_2 groups can be described by the function $\theta = 2 \, (\gamma + 0.5 \, \beta)$, where $\theta = 0°$ for the antarafacial partner at $R_{12} = \infty$ and $\theta = 180°$ after a full rotation to form cyclobutane.

The major constraints in the geometry optimization were that the CH bond lengths were kept at $1.08\,\text{Å}$ and that the line R_a joining the two ethylene centers be kept perpendicular to C_1C_4 and C_2C_3. By the conditions of the geometry optimizations at least C_2 symmetry is kept throughout the reaction.

The same parameters were employed for the $(2 \, \pi_s + 2 \, \pi_s)$ path, but the starting point of the optimizations for the points R_{12} were with $\alpha = 0°$.

If one respects the condition of an antarafacial – suprafacial approach as the starting point in the geometry optimizations, a barrier between the suprafacial – suprafacial approach can be shown down to about $2.21\,\text{Å}$. Since this R_{12} distance is also that found for the $(2 \, \pi_s + 2 \, \pi_s)$ transition state (see ref. 45), the two approaches can be thought of as two valleys leading to the same mountain pass on the energy hypersurface.

Several values of R_{12} have been explored in detail in order to analyse the process whereby the antarafacial CH_2 groups twist in opposite directions. The optimized geometries for different values of θ (the measure of CH_2 twist) are given in table 14 for each of these points along their corresponding total energies.

In order to visualize the passage from the $(2 \, \pi_s + 2 \, \pi_a)$ to the $(2 \, \pi_s + 2 \, \pi_s)$ path, an isoenergy contour plot has been constructed comparing the approach (R_{12}) to the CH_2 twist (θ). The plot is given in figure 27 where the contours have been drawn around the points given in table 14. The points are shown in figure 27 as the energy (kcal/mole) relative to two separated ethylenes, using the (15×15) CI results. The contours are 5 kcal/mole apart and are drawn as solid lines. The $(2 \, s + 2 \, s)$ path is drawn by a broken line and follows a valley formed by the optimum values of θ for given values of R_{12}. This path reaches a large plateau extending from about $R_{12} = 2.24\,\text{Å}$ and $\theta = 60°$ to about $R_{12} = 2.20\,\text{Å}$ and $\theta = 90°$. From this latter point the supermolecule may continue towards the $(2 \, \pi_s + 2 \, \pi_s)$ path $(\theta = 180°)$ without a barrier.

The change in energy (kcal/mole) along the reaction coordinate (R_{12}) is plotted in figure 28 for the three proposed mechanisms. The diradical approach has been plotted with the results reported in reference 8 (STO-3G, 15×15 CI), while the $(2 \, \pi_s + 2 \, \pi_s)$ curve is constructed with the point reported in reference 45 $(R_{12} = 2.21\,\text{Å})$ plus three other points which were optimized here. The $(2 \, \pi_s + 2 \, \pi_a)$ path is discontinuous since it branches into the $(2 \, \pi_s + 2 \, \pi_s)$ approach. It is thus difficult to talk of a $(2 \, \pi_s + 2 \, \pi_a)$ transition state, per se, but rather of a barrier from one high energy path to another viz. the $(2 \, \pi_s + 2 \, \pi_s)$.

In comparing the three reaction types for the ethylene dimerization reaction, using the same theoretical method i.e. STO-3G and (15×15) CI, it is clearly the diradical path which is favored energetically (figure 28). Of the two proposed concerted paths, the $(2 \, s + 2 \, s)$ path is preferred by about 20 kcal/mole.

Table 14. Geometric parameters optimized in function of R_{12} and the twist angle θ. The total energies were obtained by the STO-3G, (15×15) CI method.

	R_{12}	θ	α	β_{ant}	β_{sup}	γ_{ant}	γ_{sup}	δ_{ant}	δ_{sup}	R_{14}	R_{23}	$E_n(+154\ u.a.)$
I	∞	-	-	-	-	-	-	117	117	1.31	1.31	-0.230764
II	5.000	0	90	0	0	0	0	117	117	1.31	1.31	-0.224793
III	2.389	10	60	10	5	0	0	115	116	1.34	1.32	-0.109897
IV	2.389	20	50	20	5	0	0	113	116	1.34	1.32	-0.109305
V	2.389	30	50	30	5	0	0	112	116	1.34	1.32	-0.109156
VI	2.389	60	50	40	5	10	0	109	116	1.36	1.32	-0.088382
VII	2.389	90	45	50	10	20	0	109	115	1.39	1.36	-0.049721
VIII	2.389	180	0	10	10	85	-5	116	116	1.32	1.32	-0.258367
IX	2.289	10	60	10	5	0	0	115	116	1.34	1.32	-0.083147
X	2.250	30	50	30	10	0	0	112	115	1.40	1.38	-0.063757
XI	2.250	40	50	40	10	0	0	109	115	1.40	1.38	-0.063842
XII	2.250	50	50	50	10	0	0	109	115	1.40	1.38	-0.057668
XIII	2.250	70	50	50	10	10	0	109	115	1.42	1.38	-0.052976
XIV	2.250	90	50	50	10	20	0	109	115	1.42	1.38	-0.047825
XV	2.210	60	70	40	10	10	0	109	115	1.45	1.40	-0.054537
XVI	2.210	70	40	50	10	10	0	109	115	1.45	1.40	-0.056804
XVII	2.210	90	40	50	10	20	0	109	115	1.45	1.40	-0.058202
XVIII	2.210	130	20	50	10	40	0	109	115	1.45	1.40	-0.064076
XIX	2.210	180	0	30	30	75	-15	112	112	1.42	1.42	-0.099006
XX	2.189	10	60	10	5	0	0	115	116	1.34	1.34	-0.008022
XXI	2.189	70	40	40	20	15	0	109	113	1.45	1.40	-0.056987
XXII	2.189	90	40	50	20	20	0	109	113	1.45	1.40	-0.063821
XXIII	2.189	150	20	50	20	50	0	109	113	1.45	1.40	-0.076378
XXIV	2.150	70	40	40	20	15	0	109	113	1.45	1.40	-0.059027
XXV	2.150	90	40	50	20	20	0	109	113	1.45	1.40	-0.067510
XXVI	1.900	180	0	30	30	75	-15	112	112	1.52	1.52	-0.189507
XXVII	1.520	180	0	90	45	45	-45	109	109	1.52	1.52	-0.280310

The effect of CI on this path is pronounced, whereas the $(2 s + 2 a)$ path is effected to a much lesser extent. Thus the comparison of the activation barriers for the three approaches gives 42, 83, and 105 kcal/mole in the STO-3G, (15×15) CI method (two-step, $(2 \pi_s + 2 \pi_s)$, and $(2 \pi_s + 2 \pi_a)$ approaches, respectively).

We emphasize the importance of the CH_2 twisting angle, θ. Any effect which tends to diminish the barrier to twist will favorize the $(2 \pi_s + 2 \pi_a)$ path. Two such effects could be the steric hindrance of bulky cis substituents and the resonance forms by two highly polar substituents such as :

$$^+X = CH - CH = Y^-$$

This latter effect is pronounced in highly polar solvents.

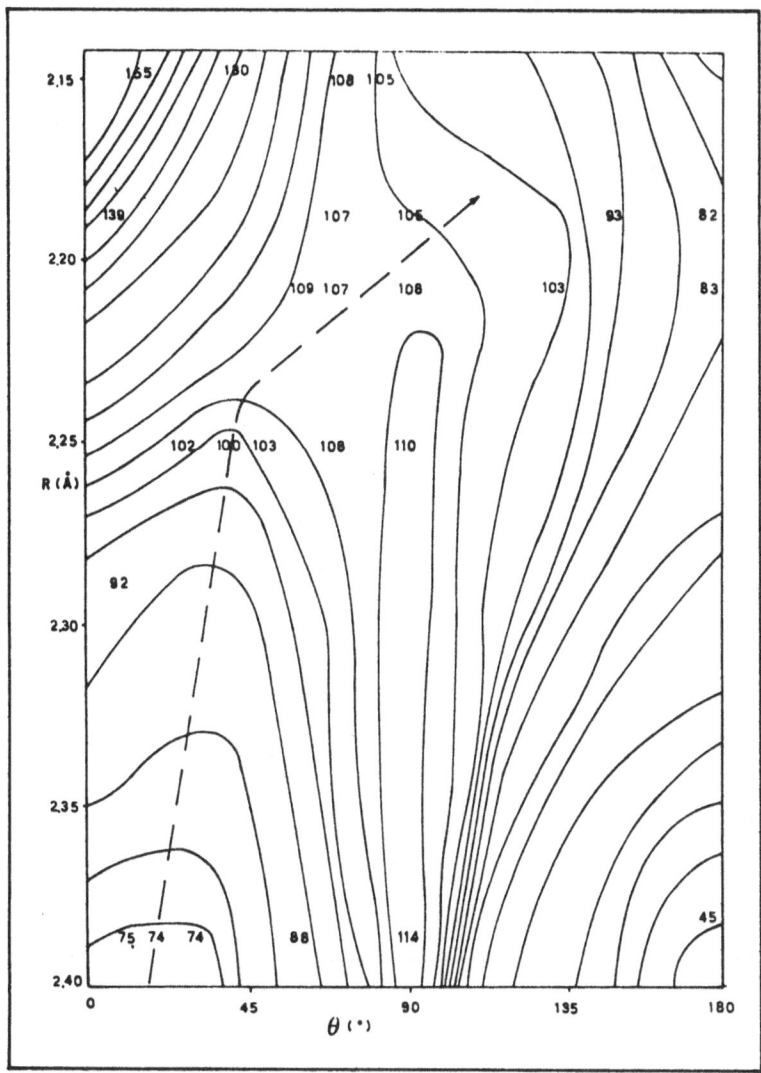

Figure 27. Isoenergetic contour drawn around the points in table 14. The abscissa, θ is the antarafacial CH_2 twist angle and the ordinate is R_{12} (= R_{34}). Each contour presents 5 kcal/ mole difference and the numbers give the energy of a particular point relative to two separated ethylenes (in STO-3G, 15 x 15 CI). The broken line indicates the ($2 \pi_s + 2 \pi_a$) pathway.

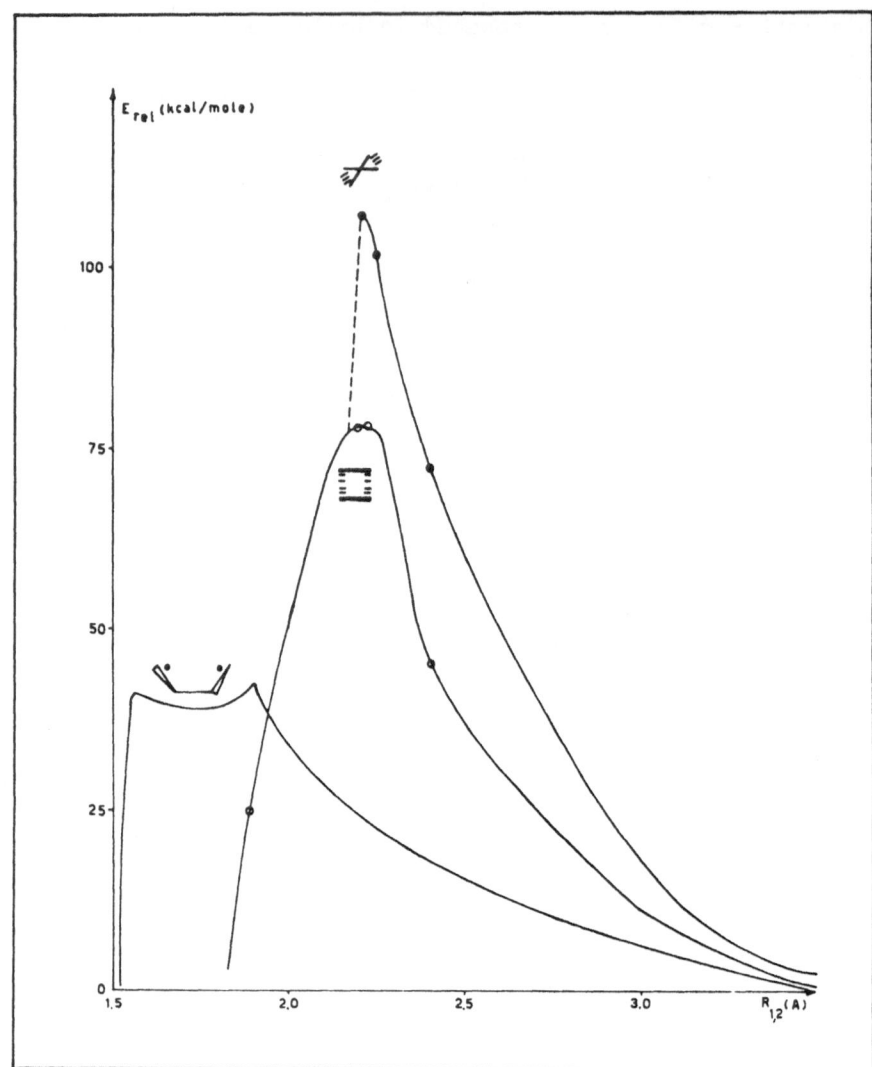

Figure 28. The energy (kcal/mole) relative to two separated ethylenes
is given along the common reaction coordinate (R_{12}) for the
three types of approach, diradical, $(2 \pi_s + 2 \pi_s)$, and
$(2 \pi_s + 2 \pi_a)$. The broken line represents the branching of
the $(2 \pi_s + 2 \pi_a)$ path into the $(2 \pi_s + 2 \pi_s)$. The circles
represent the optimized points for R_{12} in table 14 (filled
for $(2 \pi_s + 2 \pi_a)$, empty for $(2 \pi_s + 2 \pi_s)$).

4. CONCLUSION

The results described in this chapter stem from the last three
years of research carried out on different types of addition reactions.
Their heterogeneous character comes not only from the diversity of the
reactions studied but also from the rapid evolution in our means of
calculation during this period. That is the reason why our first works
on 1,3-dipolar cycloadditions were carried out with STO-3G basis set
without configuration interaction. Several points on these reaction
paths were recalculated in the slightly better, but still minimal, 7s-
3p basis set (contracted to 2s-1p) with the geometries obtained with
the STO-3G basis set. It was later on that the slightly extended, 4-31G
basis set was used in place of the 7s-3p, and then the limited CI were
carried out with STO-3G results as the starting point of the calculations. In our most recent calculations the reaction paths are obtained
employing geometry optimizations with a (55 x 55) CI. We have also been
progressively increasing the number of geometric parameters defining
the supersystem. Very fortunately, we find that these improvements that
can be made on our calculations do not fundamentally question our first
results. In particular, the CI calculations do not grossly modify the
structure of the activated complexes and the activation barriers provided by these calculations vary in a manner parallel to those obtained
by the SCF, minimal basis set method (according to the expected precision).

On the whole, the heterogeneous character of our works is only su-
perficial. They lead to a qualitatively reliable body of STO-3G results
on a homogeneous series of intra - and intermolecular addition reactions.
Of course, each type of addition constitutes a case in itself and provides particular conclusions. We recall here only the essential ones.

- *In their concerted channels the cycloadditions of linear dipoles
  to olefins are practically synchronous reactions, passing through
  planar activated complexes which are geometrically and electro-
  nically similar to the starting molecules. In the case of bent
  dipoles, the activated complex is not planar and its envelope
  angle lies between 60 and 90°.*
  *Each 1,3-dipolar cycloaddition results in a cyclic displacement
  of bonds the direction of which is determined by the more negative of the terminal atomes on the dipole.*
  *The direction of charge transfer at the transition state permits
  a prediction of the effect of substituents on the reactivity of
  the dipole and dipolarophile.*
- *Because ring-chain isomerizations present the formation of only
  one σ bond, along the reaction path, one can not really talk of
  a concerted reaction. The essential factor in the azido-tetrazole isomerization resides in the availability of the N_1 lone pair.
  The reaction paths can be broken down to several phases whose
  geometric, electronic, and energetic characteristics have been
  given. Our study leads to a correct representation of the electronic migrations which could not be foreseen by experimentalists
  due to the many conflicting possibilities of "arrows and elec-*

tron pushing" presented by azide cyclizations. This representation possesses evident characters for prediction. Any substituent which increases the availability of the N_1 lone pair will favorize the cyclization reaction. Inversely, the disparition of this lone pair by protonation increases the activation energy for the closing of the ring. We have furthermore established the tight resemblance between the cyclization of the protonated form of azido-azomethine and that of vinyl azide.

- *Finally, for the symmetric cycloadditions we have demonstrated factors which can affect the characteristics of the activated complex and we have analysed the possibility of the $(2\pi_s + 2\pi_a)$ pathway in 2 + 2 reactions.*

We propose to continue our theoretical study of reaction mechanisms in the sense of a refinement and generalization of our first results. In particular we hope to have available potential surfaces that are precise enough to be able to start calculations of the chemical dynamics and thus verify the reaction pathway found.

ACKNOWLEDGEMENTS

The authors would like to express their acknowledgements to Professor R. Daudel who gave them the opportunity to present this report at the International Symposium on Chemical Reactivity organized in Paris in 1976 and 1977. They also wish to acknowledge support for this work by a NATO grant.

REFERENCES

(1)   W.J.Hehre, W.A.Lathan, R.Ditchfield, M.D.Newton, J.A.Pople: "GAUSSIAN-70", Q.C.P.E., 11, 236 (1973).
(2)   W.J.Hehre, R.F.Stewart, J.A.Pople: J. Chem. Phys., 51, 2657 (1969).
(3)   R.Ditchfield, W.J.Hehre, J.A.Pople: J. Chem. Phys., 54, 724 (1971).
(4)   S.J.Boys: Rev. Mod. Phys., 32, 296 (1960).
(5)   D.Peeters: "BOYLOC", Q.C.P.E., 11, 330 (1977).
(6)   R.Bonaccorsi: Theoret. Chim. Acta, 21, 17 (1971).
(7)   D.Peeters, M.Sana: "ELPOT", 11, 360 (1978).
(8)   G.A.Segal: J. Amer. Chem. Soc., 96, 7892 (1974).
(9)   R.Huisgen: Angew. Chem. Int. Ed. Engl., 2, 565 (1963).
(10a)   R.A.Firestone: J. Org. Chem., 33, 2285 (1968);
(10b)   R.A.Firestone: J. Org. Chem., A, 1570 (1970);
(10c)   R.A.Firestone: J. Org. Chem., 37, 2181 (1972).
(11a)   R.Huisgen: J. Org. Chem., 33, 2291 (1968);
(11b)   R.Huisgen: J. Org. Chem., 41, 403 (1976).
(12a)   D.Popinger: J. Amer. Chem. Soc., 97, 7486 (1975);
(12b)   D.Popinger: Aust. J. Chem., 29, 465 (1976);
(12c)   D.Popinger: Private communication.
(13)   G.Leroy, M.Sana: Tetrahedron, 31, 2091 (1975); Tetrahedron, 32, 709 (1976).
(14)   J.D.Cox, G.Pilcher: "Thermochemistry in organic and organometallic compounds", Academic Press, p. 577, 593 (1970).

(15) S.W.Benson, F.R.Crickshank, D.M.Golden, G.R.Maugen, H.E.O'Neal,
      A.S.Rodgers, R.Shaw, R.Walsh: Chem. Rev., 69, 279 (1969).
(16) R.Huisgen: Private communication.
(17) C.K.Jhonson: "ORTEP", Oak Ridge National Laboratory, 3794 (1965).
(18a) F.S.Bridson-Jones, G.D.Buckley, L.H.Cross, A.P.Driver:
      J. Chem. Soc., 2999 (1951);
(18b) F.S.Bridson-Jones, G.D.Buckley: J. Chem. Soc., 3009 (1951);
(18c) G.D.Buckley, W.J.Levy: J. Chem. Soc., 3016 (1951).
(19) G.Leroy, M.Sana: Tetrahedron, 32, 1379 (1976).
(20) G.Leroy, M.T.Nguyen, M.Sana: Tetrahedron, 32, 1529 (1976).
(21a) L.I.Smith: Chem. Rev., 23, 193 (1938);
(21b) J.Hamer, A.Macaluso: Chem. Rev., 64, 473 (1964);
(21c) G.R.Delpierre, M.Lamchem: Quest. Rev., 329 (1965);
(21d) D.St.C.Black, R.F.Crozier, V.C.Davis: Synthesis, 205 (1975);
(21e) G.Leroy, M.T.Nguyen, M.Sana: Tetrahedron, 34, 2459 (1978).
(22) Y.Delugeard, J.L.Baudour, J.C.Messager:
      Cryst. Struct. Com., 3, 397 (1974).
(23) K.Folting, W.N.Lipscomb, B.Jersley: Acta Cryst., 17, 1263 (1964).
(24) K.Kubota, M.Yamalawa: Bull. Chem. Soc. Japan, 36, 1564 (1963).
(25) R.Grée, F.Tonnart, P.Carrié: Bull. Soc. Chim. France, 1325 (1975).
(26) K.N.Houk, J.Sims, R.E.Duke, R.W.Strozier, J.K.George:
      J. Amer. Chem. Soc., 95, 7287 (1973).
(27) W.J.Jennings, S.D.Worley: Tetrahedron Lett., 1435 (1977).
(28) R.Grée, G.Tonnard, R.Carrié: Tetrahedron, 32, 675 (1976).
(29) R.Huisgen, H.Seidl, I.Brüning: Chem. Ber., 102, 1102 (1969).
(30) L.A.Burke, J.Elguero, G.Leroy, M.Sana:
      J. Amer. Chem. Soc., 98, 1685 (1976).
(31a) G.Smolinsky: J. Org. Chem., 27, 3557 (1962);
(31b) J.S.Meek, J.S.Fowler: J. Amer. Chem. Soc., 89, 1967 (1967);
(31c) F.P.Woerner, H.Reimlinger: Chem. Ber., 103, 1908 (1970);
(31d) G.Beek, D.Günther: Chem. Rev., 106, 2758 (1973);
(31e) L.A.Burke, G.Leroy, M.T.Nguyen, M.Sana:
      J. Amer. Chem. Soc., 100, 3668 (1978).
(32) G.L'Abbe, G.Mathijs: J. Org. Chem., 39, 1778 (1974).
(33a) D.W.Adamson, J.Kenner: J. Chem. Soc., 286 (1935);
(33b) C.D.Hurt, S.C.Lui: J. Amer. Chem. Soc., 57, 2656 (1935);
(33c) G.L.Closs, W.Böll: Angew. Chem., 75, 640 (1963);
(33d) R.K.Barlett, T.S.Stevens: J. Chem. Soc., C, 1964 (1967);
(33e) I.Tabushi, K.Takagi, R.Oda: Tetrahedron Lett., 31, 2075 (1964);
(33f) A.Lewita, D.Parry: J. Chem. Soc., B, 41 (1967);
(33g) J.L.Brewbaker, H.Hart: J. Amer. Chem. Soc., 91, 711 (1969);
(33h) G.Tennant, R.J.S.Vevers: Chem. Comm., 671 (1974).
(34) W.D.Krugh, L.P.Gond: J. Mol. Spectroscopy, 49, 423 (1974).
(35) R.A.Henry, W.G.Finnegan, E.Lieber:
      J. Amer. Chem. Soc., 77, 2264 (1955).
(36) K.Hsu, R.J.Buenker, S.D.Peyerimhoff:
      J. Amer. Chem. Soc., 93, 2117 (1971);
      J. Amer. Chem. Soc., 94, 5639 (1972).
(37) K.Van der Meer, J.F.C.Mulder: Theoret. Chim. Acta, 37, 159 (1975).
(38) L.A.Burke, G.Leroy, M.Sana: Theoret. Chim. Acta, 40, 313 (1975).

(39) L.A.Burke, G.Leroy: Theoret. Chim. Acta, 44, 219 (1977).

(40) R.E.Townshend, G.Ramunni, G.A.Segal, W.J.Hehre, L.Salem:
      J. Amer. Chem. Soc., 98, 2190 (1976).

(41) M.V.Basilevsky, V.A.Tikhomirov, I.E.Chlenov:
      Theoret. Chim. Acta, 23, 75 (1971).

(42) F.Kern, W.D.Walters: J. Amer. Chem. Soc., 75, 6196 (1953).

(43) R.Huisgen: Acc. Chem. Res., 10, 117 (1977).

(44) R.B.Woodward, R.Hoffmann:
      "The conservation of orbital symmetry",
      Verlag Chemie - Academic Press, Weinheim (1970).

(45) J.S.Wright, L.Salem: J. Amer. Chem. Soc., 94, 322 (1972).

(46) L.Salem, C.Leforestier, G.A.Segal, R.Wetmore:
      J. Amer. Chem. Soc., 97, 479 (1975).

(47) K.Kraft, G.Koltzenburg: Tetrahedron Lett., 4357 (1967);
      Tetrahedron Lett., 4723 (1967).

(48) R.W.Holder: J. Chem. Educ., 53, 81 (1976).

STRUCTURE AND REACTIVITY: AN EXTENDED HUCKEL APPROACH

Angelo GAVEZZOTTI and Massimo SIMONETTA
Istituto di Chimica Fisica e Centro CNR,
Università di Milano, Milano, Italy

An account of some results obtained by Extended Huckel
Molecular Orbital theory is given. These include calculations of molecular structures and electronic properties,
of reaction potential energy surfaces for a few sample
systems, as well as studies on the geometry and energetics
of the chemisorption of acetylene and small fragments on
the Pt(111) surface.

Theoretical conformational analysis and the prediction of molecular
shapes dispensing with awkward experiments have become reality in
recent times. Empirical force field methods and rigorous quantum
mechanical calculations are the extremes of a wide spectrum of
approaches to this problem, differing in the amount of computing
times and of empirical parameterization. When a chemical reaction
occurs, however, bonds are broken and formed, so that force field
methods are usually unable to describe correctly the properties of
molecular systems far from equilibrium; one has to resort to quantum
mechanics to obtain reaction surfaces. Even so, although the reaction energetics may be calculated, dynamical factors are still likely
to play a major role in the determination of reaction paths.

Extended Huckel Theory (EHT) is a largely empirically parameterized quantum mechanical method of calculating molecular properties. The Fock operator is simplified by neglecting all exchange
and coulomb terms; the H part is an unspecified operator, whose
matrix elements, as resulting from the usual LCAO-MO formulation

$$\varphi_i = \sum_p c_{pi} \chi_p$$

are approximated by the formula

$$H_{pq} = 1/2 \, K \, (H_{pp} + H_{qq}) \, S_{pq}$$

145

R. Daudel, A. Pullman, L. Salem, and A. Veillard (eds.), Quantum Theory of Chemical Reactions, Volume I, 145–159.
Copyright © 1979 by D. Reidel Publishing Company.

K is a constant empirically adjusted to about 1.75, and the H_{pp}'s
are approximated by atomic valence ionization potentials.
Overlap matrix elements (S_{pq}) are rigorously computed, and variation
with respect to the coefficients of the LCAO expansion leads to
the secular equation

$$(H - \epsilon_i S)c_i = 0$$

with the orthogonality relationship

$$\int \varphi_i \varphi_j \, d\tau = c_i Sc_j^T$$

The total energy is computed by the formula

$$E = 2 \sum_i \epsilon_i$$

If one compares with the complete total energy expression (for a
closed shell)

$$E = 2 \sum_i \epsilon_i - \sum_{i,j} (2J_{ij} - K_{ij}) + e^2 \sum Z_A Z_B / R_{AB}$$

one can see that the performance of the approximation lies in the
cancellation of the second and third terms in the right-hand side
of the above expression.

Since its birth in 1963 (Hoffmann, 1963) EHT has met with
harsh criticism (Dewar,1971), as well as with increasing popularity
with chemists; attempts to rationalize the reasons of its success
or failure have appeared (Blyholder and Coulson,1968; Goodisman,
1969). Originally proposed in the context of organic covalent
compounds (Hoffmann,1963; 1964), its scope is now widening, and,
through the link of coordination chemistry (see, e.g., Summerville
and Hoffmann, 1976), attention is now focusing on its applications
to the fascinating, and in many respects baffling, phenomena of
chemisorption and surface processes (see, for a review, Gavezzotti
and Simonetta,to be published). Recently (Lohr and Pyykko, 1979)
a relativistically parameterized Extended Huckel theory has been
formulated. The method should be able to provide a semi-quantitative
description of relativistic effects in chemical bonding when one
or more atoms with large nuclear charge are present, and should
prove to be particularly useful in the study of transition metal
clusters.

Although the study of bending, rather than stretching motions,
is EHT's best performance, appropriate precautions make it possible
to study reaction surfaces as well. The following is a brief account
of our experience in the use of EHT in various structure and reactivity problems.

THE MO's OF BRIDGED ANNULENES

EHT is a valence orbital method: it takes explicitly into account all the electrons in the valence shell of the atoms in a molecule. As an example of the advantages it offers over simple Huckel treatments, the case of bridged annulenes can be cited. These are a very uncommon class of compounds (Vogel,1968) in which the presence of bridges connecting the various nodes of the annulene chain imposes a distortion from planarity to the annulene ring (see Figure 1). As a consequence, the π system is no longer orthogonal to the σ frame, whose contributions to the former π orbitals

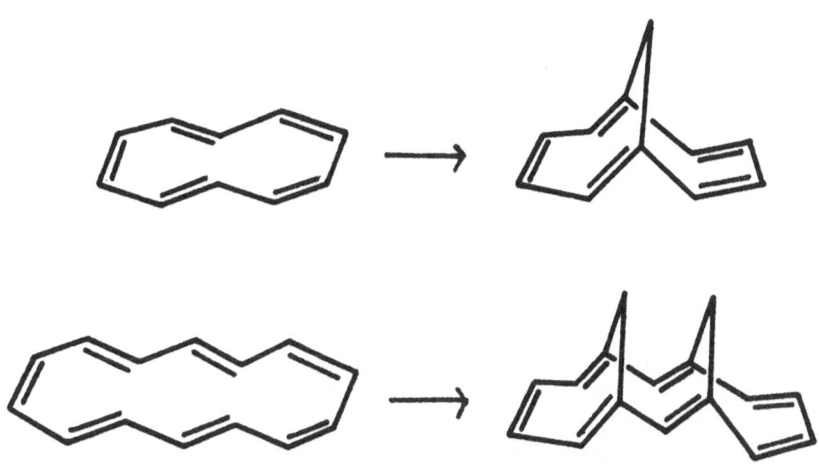

Figure 1. Distortion of the [10]annulene and [14]annulene rings imposed by the presence of the bridges.

result in significant shifts of the MO energies. In a series of calculations on substituted bridged annulenes (Gavezzotti and Simonetta,1976) we have shown that EHT gives a reasonable description of this σ/π mixing, which is essential to the discussion of level ordering in relationship with photoelectron spectra and of the aromaticity of these compounds.

Molecular shapes and reactivity of some members of the annulene family are well known from experiment; EHT predictions on the structure of the bridge in syn-bismethano[14]annulene, and the calculated atomic charges along the annulenic perimeter, were in good correspondence with X-ray diffraction structural data and with relative reactivities to electrophilic substitution, respecti-

vely. Finally, certain details of the relative stability of the
annulenic and bisnorcaradienic forms in [10]annulene derivatives
were succesfully explained:

REACTIVITY OF HALOACETYLENES TO SULPHUR NUCLEOPHILES

The availability of experimental data on the reaction

$$RC \equiv CX \; + \; R'S^- \longrightarrow RC \equiv CSR' \; + \; X^-$$

prompted an examination of the EHT predictions concerning the
geometry of attack (α, β or to the halogen) by the nucleophile,
and the relative ease of reaction for the various halogens and
for R = methyl or phenyl (Beltrame, Gavezzotti and Simonetta,1974).
Parameter choice and optimization were major concerns in tackling
the problem; reasonable dissociation curves for halogenophenyl
and -methyl acetylenes were finally established. Sections through
the reaction potential hypersurface allowed the identification of
some stable intermediates, as well as the discussion of the favoured
attack geometries; for instance, it was concluded that α-attack
follows the order of reactivity $F > Cl > Br > I$, and phenylderivatives
were found to be more prone to reaction than methylderivatives.
Our experience here more than confirms the warning that extensive
parameter optimization is to be carried out before attempting
calculations on molecules containing highly electronegative atoms.

A SIGMATROPIC SHIFT AND AN ELECTROCYCLIC REACTION

The explicit inclusion (and the prevailing effect) of overlap
in EHT calculations make them the natural tool for studies along
the lines of orbital symmetry conservation rules. There are however
cases in which steric requirements are in antagonism with the purely
electronic factors that would make a reaction symmetry-allowed or
-forbidden. This is the case in the highly strained bicycloheptenes

(see Figure 2) where the symmetry-allowed antarafacial sigmatropic
shift cannot take place, and inversion of configuration (required
by maximum overlap) may be hindered by the presence of bulky
substituents. Our calculations (Gavezzotti and Simonetta,1975a)

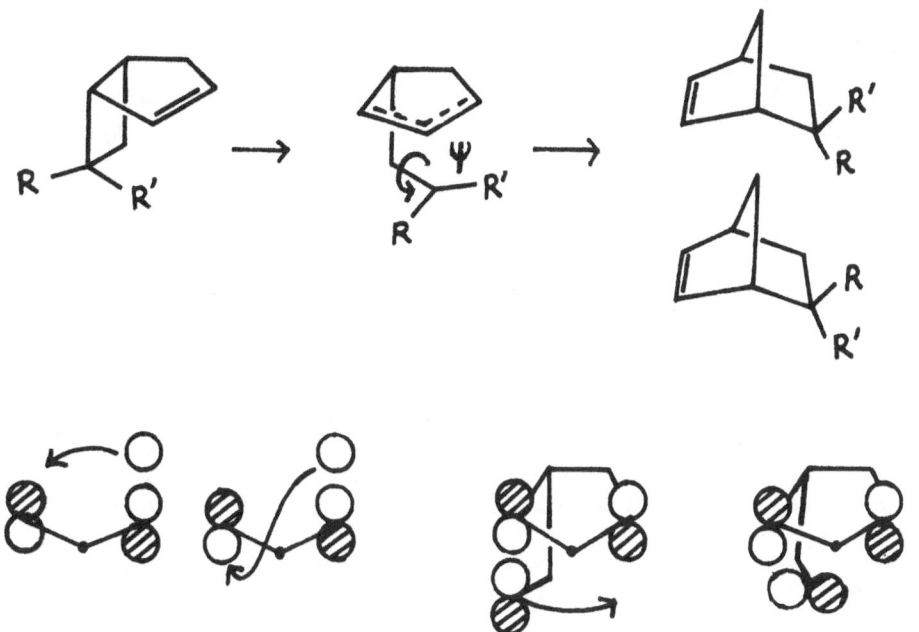

Figure 2. Sigmatropic shift in bicycloheptene derivatives.
Below: forbidden suprafacial and allowed antarafacial shift
for allyl + H; forbidden (without) and allowed (with inversion of configuration) suprafacial shift in bicycloheptene.

for this reaction system (with R = R' = H or R = H and R' = methyl)
yielded reaction energy surfaces as a function of the displacement
of the migrating group and of the ψ rotation (Figure 2). Essentially
it was shown that in such extreme a situation the effects of faster
growth (slower decrease) of orbital overlap was evident only in
the slopes of the potential energy hill, whose height was mainly
assessed by strain factors. In particular, for the methyl derivative all paths that resulted in pushing the substituent towards
the five-membered ring, as a consequence of inversion of configuration at the migrating group, were found to be unfavourable. A
detailed comparison was succesfully conducted with the results of
kinetic experiments, available through a series of elegant and
accurate experimental works (Berson, 1972).

Ring opening and closure in substituted cyclopropanes falls
into the class of electrocyclic reactions. There are precise rules
that govern the allowedness or forbiddenness of such reactions,
but the presence of substituents influences them in two ways: by
changing the level ordering of the MO's and their symmetry, and
by possibly making the reaction a non-concerted one (i.e., stabiliz-
ing diradical-type intermediates). For the molecules:

EHT calculations were done (Gavezzotti and Simonetta,1975b; Simonetta
and Gavezzotti,1976a,b) in order to assess the relative importance
of bond breaking and successive rotation or rotations at the methylene groups. In this case too, very accurate experimental rate data
were available (Doering and Sachdev,1974; Berson, Pedersen and
Carpenter,1976, and references therein). A few general principles
were detected or confirmed by EHT calculations: the ease of rupture
of the bond joining the most substituted carbon atoms was one of
them. Again, detailed comparison of relative activation energies
with rate data yielded a fair agreement, thus encouraging the
preliminary study of reaction mechanisms on the basis of energy
data only, although a full discussion would require the inclusion
of dynamical factors as well.

CHARGED SPECIES

Large charge separations are not welcome when the use of EHT
is planned; stretching energies in charged species, as computed
by EHT, are less and less accurate, as the difference in electrone-
gativity between the atoms joined by the bond increases. The
situation is less uncomfortable when bending motions in ionic
molecules are considered; thus EHT gave results comparable to
ab initio ones for the barrier to the symmetric inversion of sulfonium, H_3S^+ (Gavezzotti and Simonetta, in the press (a)). Given
this succesful attempt, the barriers were calculated also for the
methyl- and phenyl-susbtituted derivatives; the results are summarized in Table 1.

An EHT potential energy surface was constructed (Gavezzotti
and Simonetta,1977a) for the reaction

TABLE 1

Energy barriers (Kcal/mole) for the symmetric
inversion in sulfonium compounds

	equilibrium HSH or CSC angle	EHT barrier	ab initio barrier
H_3S^+	93°	25	30.4[a]
$(CH_3)_3S^+$	109°	24	-
$(C_6H_5)_3S^+$	~120°	~0	-

[a]R.E.Kari and I.G.Csizmadia, Can.J.Chem. 53,3747 (1975)

in which two neighbour molecules in the crystal trade a methyl
group in an intermolecular fashion, the reaction rate being largely
enhanced in the solid state with respect to solution by the favourable packing arrangement (Sukenik, Bonapace, Mandel, Lau, Wood and
Bergman, 1977). Activation energies for bond breaking were computed
to be very small, of the order of a few Kcal/mole, but this is not
surprising, since the EHT description of polar bonds is poor. Trends
in atomic charges and bond overlap populations are probably more
reliable; in this case, this information was used to attack the
problem of electrostatic energy in the crystal and to rationalize
some aspects of the reaction mechanism, that leads to a stable
ion-pair intermediate. One most interesting energetic aspect of
this reaction is the ease with which the reaction-promoting molecular displacements can occur in the crystal; the potential energy
walls along these directions around the molecule that undergoes
reaction were calculated to be especially "soft" ones by means of
non-bonded intermolecular potentials.

CHEMISORPTION AND SURFACE CHEMISTRY

The breakthrough in the use of EHT in heavy-metal chemistry
came with a number of papers by Hoffmann and coworkers (Rosch and
Hoffmann,1974; Hay, Thibeault and Hoffmann,1975), where many
delicate points in coordination chemistry were succesfully discussed. About at the same time, the idea of modeling a metal surface
by a cluster of metal atoms, and an adsorbate layer by a single
molecule, to study chemisorption, was proposed (Fassaert, Verbeek
and Der Avoird,1972; Anders, Hansen and Bartell,1973; Anderson and
Hoffmann,1974). In this field, the competition with more rigorous
quantum mechanical calculations is less severe, due to present-time
unability of ab initio methods to deal with such a prohibitive

amount of calculations. The need for a cheap but reliable approximate scheme was therefore particularly felt. This subject has been
reviewed in many occasions (see e.g. Messmer,1978; Gavezzotti and
Simonetta, to be published).

In our laboratory, much work was dedicated to the problem of
acetylene chemisorption on Pt(111) from a theoretical standpoint.
This system has been the subject of extensive and tenacious
experimental work, whose conclusions are still subject to criticism
and discussion; the main features that emerge from LEED (Kesmodel,
Dubois and Somorjai,1978), UPS (Demuth,1979; Lo, Chung, Kesmodel,
Stair and Somorjai,1977), EELS (Ibach, Hopster and Sexton,1977)
and TPTD (Demuth,1979) studies are summarized in the following
scheme:

HC≡CH + Pt(metal) ⟶   H⟍C≡C⟋H   Low T to room T
                        I   ┊   "metastable" phase
                            ⁄⁄⁄⁄

Intermediate
T range (just
above room T)

II   H⟍C⟋H   ⟶   III   H H H   ⟶   IV   H⟍C⟋H H
          ┊   ⟍C⟋   ┊ ⟍
          C   ┊   C—H
         ⁄⁄⁄⁄   C   ┊
                              ┊   ⁄⁄⁄⁄
                             ⁄⁄⁄⁄

————————————————————————————
                    ↓
        fragmentation (high temperature)

The essential questions to answer are: a) which is the molecular
species adsorbed at each temperature, and possibly what are its
dimensions; b) which is the particular surface site chosen by each
species; c) what is the binding energy for each species on the
surface; d) what is the mechanism by which the species transform
one into another. Besides the information shown in the scheme,
experimental results provide arguments in favor of "on top" site
for I, and of triangular site for II and III. Back-bent acetylene
is at a distance of 2.4 Å above the surface.C-C and Pt-C distances
of 1.5 and 2.0 Å respectively were determined for III on the trian

gular site (Kesmodel, Dubois and Somorjai,1978). No quantitative
answer has yet been presented to questions c) or d).

From the theoretical standpoint, the first problem in the use
of EHT is posed by parameterization. There is some uncertainty as
to which are the best values for the ionization potentials and the
orbital exponents for the crucial Pt d functions; there is even
some uncertainty as to which is actually the valence shell of Pt
(5p,5d,6s or 5d,6s,6p). We anticipate here the conclusions relative
to these points that we have reached after many experiences (Gavezzotti and Simonetta,1977b; 1979; in the press (b)): i) 5d,6s,6p
is the best choice for the Pt valence shell, but 5p orbitals
contribute a sort of inner-shell repulsion that can be approximated
by empirical formulas such as (Anders, Hansen and Bartell,1973):

$$E(repulsion) = A \exp(-B R_{Pt-C})$$

whose parameters are to be obtained from test calculations with
and without 5p orbitals; ii) the contracted, double-zeta functions
are to be preferred over diffuse, single-zeta ones for 5d orbitals.
The use of diffuse functions fails to reproduce bonding effects
if the inner-shell repulsion is included, and favours on-top Pt-C
interactions if repulsion is left out; a correct balance is obtained
on the contrary between the large binding effect brought about by
the use of double-zeta functions and the inner-shell 5p repulsion,
yielding reasonable binding energies and favouring adsorption on
a threefold site. All this is illustrated by the examples shown
in Figure 3. The EHT parameters are collected in Table 2.

Table 3 collects a number of adsorption energies for acetylene
and various fragments. From this, it is evident that the answer
EHT gives to question a) is roughly in agreement with the results
presented in the scheme. Loosely bound, easily formed species
such as back-bent acetylene exist at lower temperatures; II and
III are appropriate to moderate temperature ranges; highly frag-
mented species pertain to the high temperature domain. The answer
to question b) is in beautiful agreement with experiment when the
more reliable double-zeta functions are used. In what concerns
binding energies, although the absolute values are highly sensitive
to the choice of the basis set, the order acetylene$<$ II $<$ III is
not. In particular the difference between the adsorption BE's of
II and III is the same with the two different basis sets.

The answer to question d) is far beyond the scope of present
day experimental techniques. It is in this field that theoretical
calculations can be most useful. For the reaction I$\longrightarrow$ II the
values collected in Table IV have been calculated. It can be seen
that, despite the differences in absolute values, due mainly to
the overestimation of adsorption BE's with the single-zeta

no-repulsion basis set, the relative trends are once more basis
set independent. The metal offers a catalytic activity that lowers
the activation energy of the process; the transformation of I into
II was hypotized to take place via a transition state whose
structure has been determined by ab initio calculations (Dykstra
and Schaefer,1978). The transition state is stabilized by adsorption
more than the reagent is; the same is true for the product II,
so that also the reaction enthalpy is less positive over a metal
than in the gas phase.

The reaction II + H → III is calculated to be a facile and
highly hexothermic one (as could be expected by considering the
reverse reaction). One might conclude that, all other things being
equal, III is the preferred species on the Pt(111) surface as the
temperature is raised; this is in agreement with some of the
experimental evidence and with other EHT calculations that allowed
a comparison of observed and calculated UPS spectra (Baetzold, in
the press). Also the formation of IV should be thermodynamically
easy; EHT calculations suggest however a certain reluctance of the
C atom bound to the single H atom to find a suitable placement
on the surface. Strong Pt...H non-bonded interactions are another
argument against the existence of IV on the surface.

One final point deserves mention. The EHT energy for back-bending
of acetylene to HCC = 140°, and for stretching of the C≡C
distance to 1.30 Å (the average values found in Pt complexes) is
about 50 Kcal/mole, while the adsorption energy of this species
is only 39 Kcal/mole. For adsorption to take place at all, one
must conclude that the severely distorted conformation of the
acetylene molecule found in complexes does not apply to chemisorbed
acetylene. As a matter of fact, in a first approximation the
back-bending and/or stretching of the C≡C or even C-H bonds might
be thought of as arising from a balance between the tendency of
the molecule to remain undistorted (as measured by bending and
stretching force constants) and Pt...H non-bonded interactions,
that can be calculated by some sort of 6-exp potential. Of course,
this does not take into account electronic factors of the Pt-C
bonds. These strain energies are easily calculated for acetylene
over a triangular site; the results are presented in Table 5. As
can be seen, at the bonding distance of acetylene (≈ 2.2 Å) the
back-bend angle is calculated to be only 158°, while the C≡C
bond length is extremely reluctant to stretch. In this conformation,
the distortion energy roughly equals the adsorption binding energy.

We thank dr. R.C.Baetzold for letting us know his results prior
to publication.

TABLE 2

Extended Huckel parameters[a]

		single zeta basis set		double zeta basis set[b]	
		VOIP	zeta	VOIP	zeta
5p		-56.87	5.10	-56.87	5.10
5d		-11.39	2.31	-12.59	6.013(0.6334) 2.696(0.5513)
6s		-6.382	1.20	-9.077	2.554
6p		-6.0	1.0	-5.475	2.554

[a]See Gavezzotti and Simonetta,1977b; in the press (b). VOIP's in eV. Coefficients are given for the two functions of the double zeta expansion. [b]Summerville and Hoffmann,1976.

TABLE 3

Calculated fragment adsorption energies (Kcal/mole) and distances from the surface (Å)

		single zeta			double zeta		
	site	BE	distance		site	BE	distance
C	on top	302	0.85				
CH	on top	224	1.40				
CCH	on top	205	1.40				
H_2CC	on top	178	1.45		triang.	69	2.13
H_3CC	on top	205	1.50		triang.	97	2.06
H_3CCH	on top	115	2.50				
$H_{C\equiv C}H$[a]	di-sigma	~0	2.3		triang.	39	2.19

[a]$R(C\equiv C)$ = 1.30 Å, HCC angle 140°.

TABLE 4

Calculated activation energies and heats of reaction for the
process acetylene ⟶ III (Kcal/mole)

	single zeta		double zeta	
	$\Delta E^{\neq}$	ΔH	$\Delta E^{\neq}$	ΔH
gas phase	104	+83	104	+83
over 1 Pt atom	89	+5	-	-
over 3 Pt atoms	93	+39	74	+28
over a surface (cluster)	24	-94	65	+14

TABLE 5

Calculated geometry of the acetylene molecule as a function of
the distance, z, from the metal surface. Optimization of stretch
and bend energies plus non-bonded energy. Distances in Å, angles
in degrees.

z	HCC angle	R_{C-C}	R_{C-H}
∞ [a]	180	1.21	1.09
2.5	165	1.21	1.09
2.1	158	1.21	1.10
1.7	145	1.21	1.11
1.3	130	1.21	1.15

[a]Gas-phase structure of the acetylene molecule.

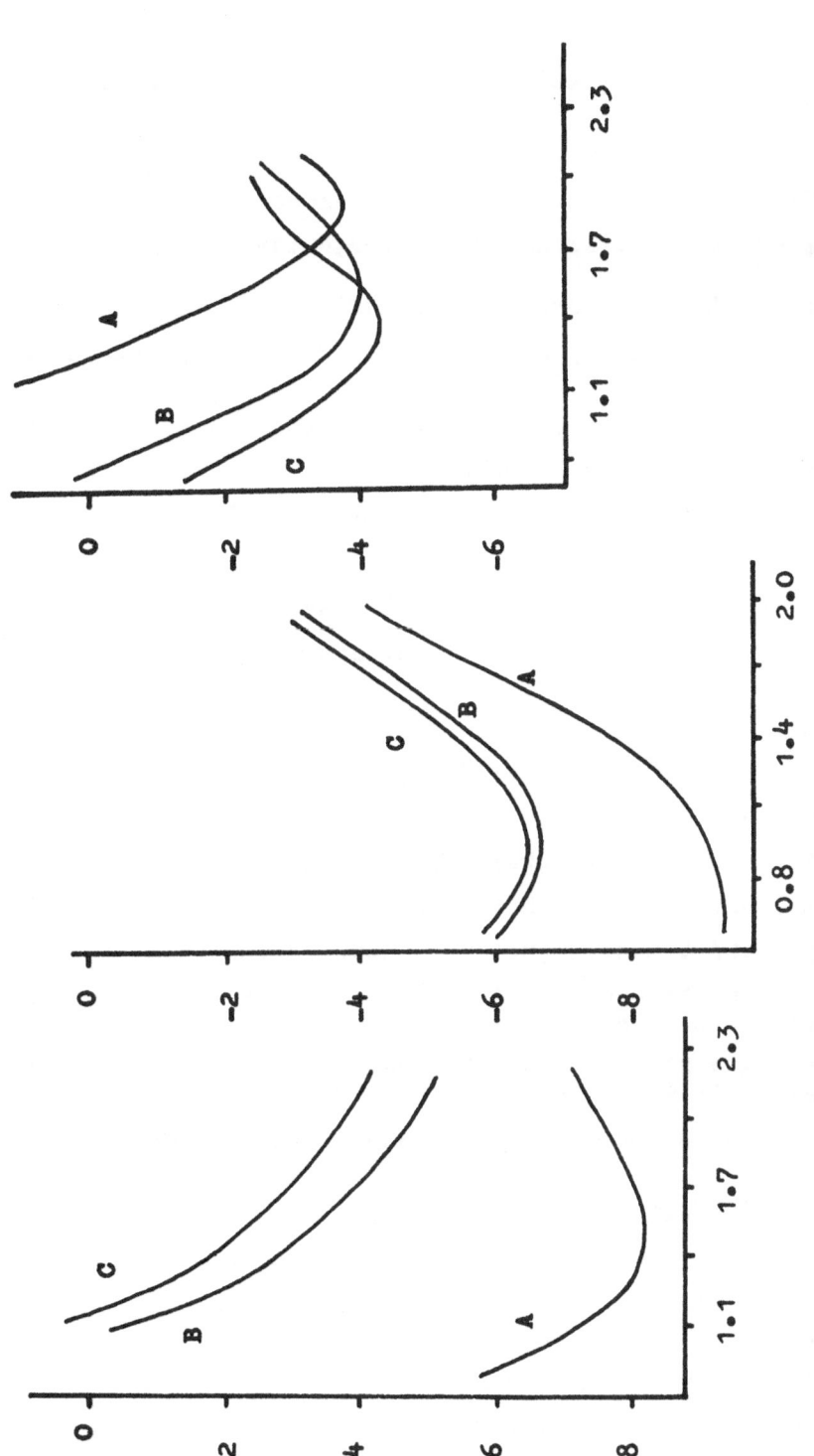

Figure 3. Adsorption binding energies (eV) for H_3CC on Pt(111). A on top, B bridging, C triangular site. Abscissa is the distance from the surface. At left, single-zeta no-repulsion basis set; center, double zeta no-repulsion; right, double zeta with repulsion.

REFERENCES

Anders,L.W., Hansen,R.S., and Bartell,L.S.:1973,J.Chem.Phys.59, p.5277.

Anderson,A.B., and Hoffmann,R.:1974,J.Chem.Phys.61,p.4545.

Baetzold,R.C.: in the press, Chemical Physics.

Beltrame,P., Gavezzotti,A., and Simonetta,M.:1974,J.Chem.Soc. Perkin II,p.502.
Berson,J.A.:1972,Accounts Chem.Res.5,p.406.

Berson,J.A., Pedersen,L.D., and Carpenter,B.K.:1976,J.Am.Chem.Soc. 98,p.122.

Blyholder,G., and Coulson,C.A.:1968,Theoret.Chim.Acta 10,p.316.

Demuth,J.E.:1979,Surface Sci.80,p.367.

Dewar,M.J.S.:1971,Topics in Current Chemistry vol.23,p.1.

Doering,W. von E., and Sachdev,K.:1974,J.Am.Chem.Soc.96,p.1168.

Dykstra,C.E., and Schaefer,H.F.III:1978,J.Am.Chem.Soc.100,p.1378.

Fassaert,D.J.M., Verbeek,H., and der Avoird,A.:1972,Surf.Sci. 29,p.501.

Gavezzotti,A., and Simonetta,M.:1975a,Tetrahedron 31,p.1611.

Gavezzotti,A., and Simonetta,M.:1975b,Tetrahedron Letters,p.4155.

Gavezzotti,A., and Simonetta,M.:1976,Helv.Chim.Acta 59,p.2984.

Gavezzotti,A., and Simonetta,M.:1977a,Nouveau J.Chim.2,p.69.

Gavezzotti,A., and Simonetta,M.:1977b,Chem.Phys.Letters 48,p.434.

Gavezzotti,A., and Simonetta,M.:1979,Chem.Phys.Letters 61,p.435.

Gavezzotti,A., and Simonetta,M.:in The Chemistry of the Sulfonium Group,P.Sterling,Ed., in the press (a).

Gavezzotti,A., and Simonetta,M.:12th Jerusalem Symposium, april 1979, Jerusalem, in the press (b).

Gavezzotti,A., and Simonetta,M.:Advances in Quantum Chemistry, to be published.

Goodisman,J.:1969,J.Am.Chem.Soc.91,p.6552.

Hay,P.J., Thibeault,J.C., and Hoffmann,R.:1975,J.Am.Chem.Soc.97, p.4884.

Hoffmann,R.:1963,J.Chem.Phys.39,1397.

Hoffmann,R.:1964,J.Chem.Phys.40,p.2474.

Ibach,H.,Hopster,M., and Sexton,B.:1977,Appl.Surf.Sci.1,p.1.

Kesmodel,L.L., Dubois,L.H., and Somorjai,G.A.:1978,Chem.Phys.
Letters 56,p.267; J.Chem.Phys.,in the press.

Lo,W.J., Chung,Y.W., Kesmodel,L.L., Stair,P.C., and Somorjai,G.A.:
1977,Solid St.Commun.22,p.335.

Lohr,L.L., and Pyykko,P.: 1979,Chem.Phys.Letters 62,p.333.

Messmer,R.P.:1978,in Semiempirical Methods of Electronic Structure
Calculations,Part B: Applications,G.A.Segal,Ed.,Plenum Press,
New York.

Rosch,N., and Hoffmann,R.:1974,Inorg.Chem.13,p.2656.

Simonetta,M., and Gavezzotti,A.:1976a,Atti Accad.Sci.Torino 111,
p.93.

Simonetta,M., and Gavezzotti,A.:1976b,in Structure and Bonding,
J.D.Dunitz,Ed., vol.27, Springer-Verlag, Berlin.

Sukenik,C.N., Bonapace,A.P., Mandel,N.S., Lau,P., Wood,G., and
Bergman,R.G.:1977,J.Am.Chem.Soc.99,p.851.

Summerville,R.H., and Hoffmann,R.:1976,J.Am.Chem.Soc.98,p.7240.

Vogel,E.:1968, in Proceedings of the Robert A.Welch Foundation
conference in Chemical Research, XII, Organic Synthesis, Houston.

SYMMETRY AND THE TRANSITION STATE

J.N.Murrell
School of Molecular Sciences, University of Sussex,
Brighton BN1 9QJ, U.K.

A review is given of the relationship between symmetry and the
geometry of transition states.

The relationship of saddle points to reactant and product valleys
(or minima) on potential energy surfaces, is the key to our under-
standing of chemical reactivity. This relationship involves both energy
and configuration. The relative energy will show itself primarily in
the enthalpy of activation of a reaction and the relative configuration
primarily in the entropy of activation.

The most important role of symmetry in molecular physics is to
provide " yes - no" answers. For example, one can deduce that a
spectroscopic transition is allowed or forbidden or a molecular configuration can, or cannot, be a transition state. It is more difficult
using symmetry to say whether an allowed transition will give a strong
or weak spectroscopic band, or whether a Jahn-Teller distortion will be
large or small. Symmetry is therefore no substitute for quantum
mechanical calculations, but it may suggest which calculations are
worth doing.

There are two quite distinct aspects of transition states which are
amenable to symmetry arguments. The first is the symmetry of the
electronic wave function of the relevant state and the second is the
symmetry of the normal modes of vibration.

Before we can use symmetry rules we must be clear on the definition
of a transition state.

Reactants and products are represented by valleys or minima on a
potential energy surface. A transition state is the maximum on a
minimal energy path between reactants and products. Necessary conditions
for a point to be a transition state are that the slope of the potential
energy function be zero in any direction and that only one of its
principal curvatures is negative. Although we have yet to define

161

R. Daudel, A. Pullman, L. Salem, and A. Veillard (eds.), Quantum Theory of Chemical Reactions, Volume I, 161-176.
Copyright © 1979 by D. Reidel Publishing Company.

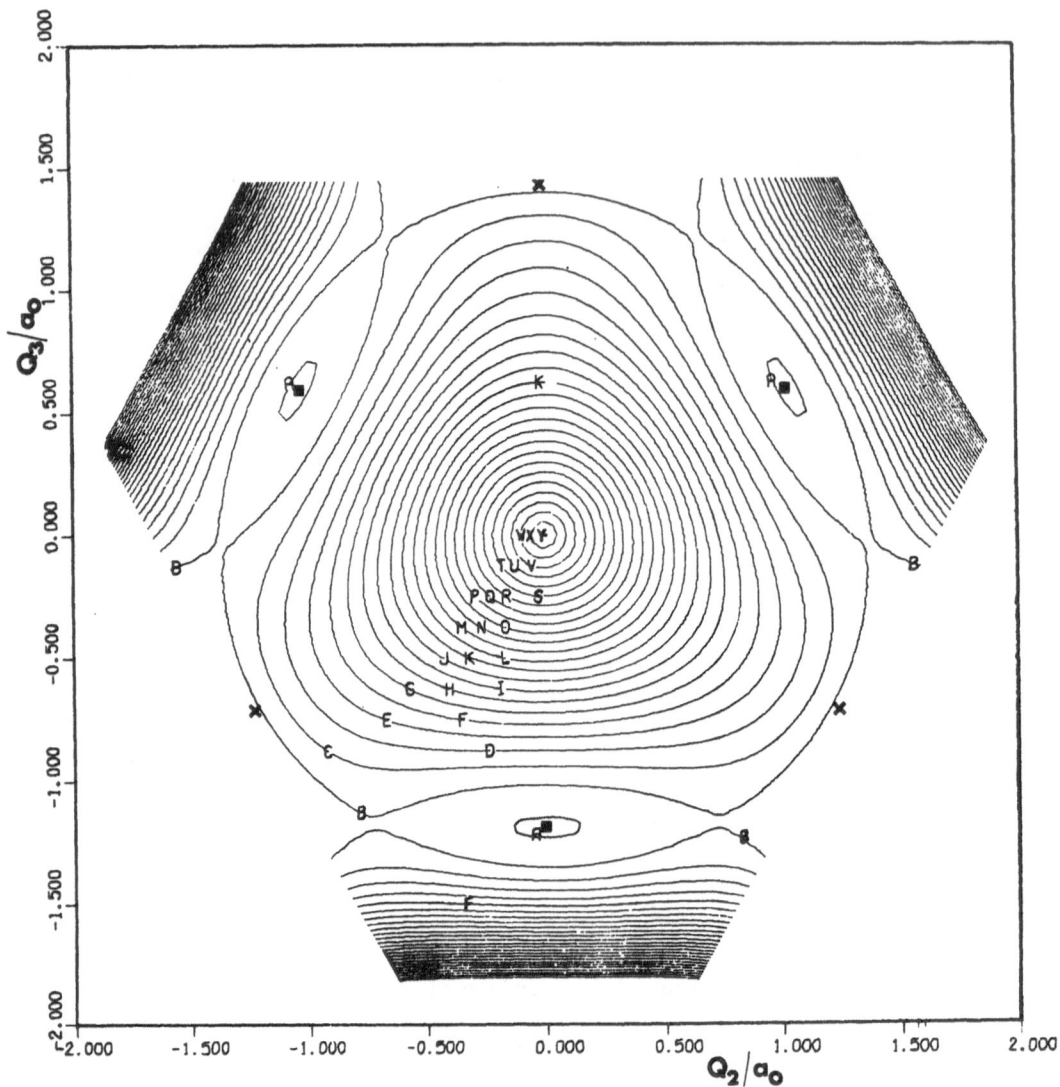

Figure 1. Contours of the potential energy surface for the ground state of H_3 for a perimeter of the triangle $7.1a_0$. Only the physical space of (Q_2, Q_3) is shown. Contour A is 0.016 a.u. and subsequent contours increase by 0.004 a.u. The collinear saddle points are indicated (X) and the points shown have C_{2V} symmetry [4].

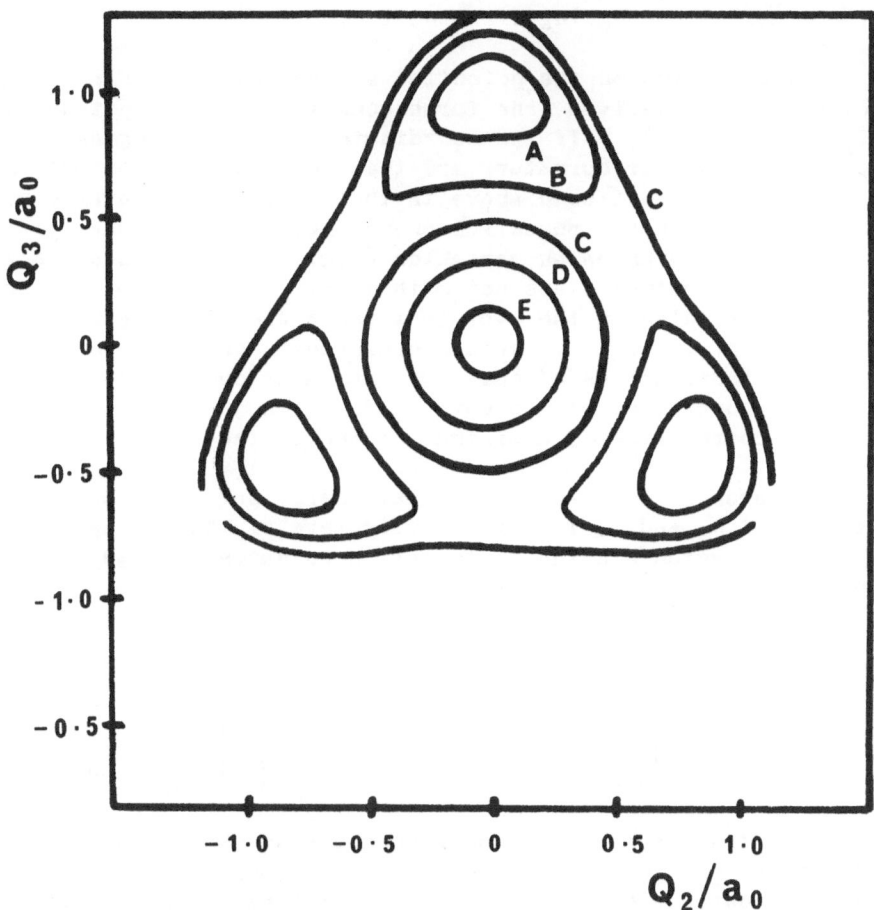

Figure 2. Contours of the potential energy surface of O_3 for a perimeter 4.709Å. Contour A is -5.75eV and subsequent contours increase by 1eV [3].

unambiguously what is meant by a principal curvature, the reason for
the second condition is obvious: if there are two directions of
negative curvature then there must be a lower energy path between one
side of the point and the other which does not pass through the point
in question [1].   Quite simply, one can go from one side of a hill to
the other without going over the top.

At any point on the potential surface one can establish the matrix
of second derivatives (the force constants) of the potential
$F_{ij} = (\partial^2 V/\partial r_i \partial r_j)$.   If the coordinates r_i are orthogonal then the
principal axes of curvature are the eigenfunctions of this matrix and
the condition mentioned above is that only one eigenvalue must be
negative.   However, potential energy surfaces are usually defined with
internal coordinates as variables (e.g. bond lengths and bond angles)
and such coordinates are not orthogonal.   The lack of orthogonality
finds expression in the fact that there are off-diagonal terms in the
kinetic matrix, T_{ij}, in such coordinates.   Under these conditions the
principal directions of curvature are most conveniently defined as the
normal coordinates of the point defined in just the same way as the
normal coordinates of a minimum of the surface.

Figures 1 and 2 show potential energy surfaces for the ground
states of H_3 and O_3 for fixed perimeters of the triangle.   These figures
have axes which are the E' normal coordinates (Q_2, Q_3) of a D_{3h} structure
defined by

$$
\begin{bmatrix} Q_1 \\ Q_2 \\ Q_3 \end{bmatrix} = \begin{bmatrix} \sqrt{\frac{1}{3}} & \sqrt{\frac{1}{3}} & \sqrt{\frac{1}{3}} \\ 0 & \sqrt{\frac{1}{2}} & -\sqrt{\frac{1}{2}} \\ \sqrt{\frac{2}{3}} & -\sqrt{\frac{1}{6}} & -\sqrt{\frac{1}{6}} \end{bmatrix} \begin{bmatrix} R_1 - R^o \\ R_2 - R^o \\ R_3 - R^o \end{bmatrix}
$$

The figures therefore show contours of the potential for a fixed
perimeter of the triangle defined by the positions of the three atoms.
The contours have been drawn from analytical functions determined by
fitting spectroscopic data or ab-initio calculations [3,4].

The centre of the two figures is the point of D_{3h} configuration and
we note that the H_3 surface is sharply peaked but the O_3 surface is
rounded.   The H_3 surface is the lower sheet of a conical intersection
(see figure 3) associated with the fact that the ground state of H_3 for
D_{3h} configurations is doubly degenerate (E').   We can apply the same
arguments, that is, those of Jahn and Teller [5], for the D_{3h} structure
not being a saddle point as could be made for it not being a true
minimum on the surface [6].   The Jahn-Teller theorem proves that there
must be a non zero first derivative of the potential at this point along
the direction of some normal coordinate, and this is inconsistent with
our definition of a transition state.

A simple MO treatment of D_{3h}, X_3 molecules, where X is a main group

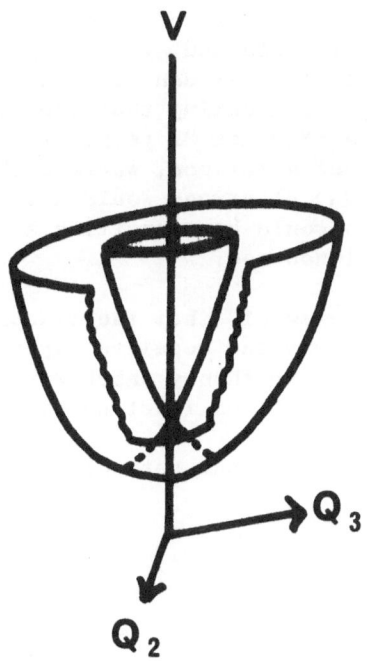

<u>Figure 3</u>.   A Jahn-Teller conical intersection. Figure 1 shows contours
only of the lower sheet of such a 2-valued surface.

element, gives the following increasing order for the valence molecular
orbital energies:

$$a_1'(s), e'(s), a_1'(p\sigma_r), a_2''(p\pi), e'(p\sigma_t), e''(p\pi)a_2'(p\sigma_t), e'(p\sigma_r)$$

where the suffix r (radial) labels $p\sigma$ orbitals which are pointing
towards the centre of the triangle and t labels $p\sigma$ orbitals at right
angles to this set (tangential). It can be seen that neutral molecules
with 3 (e.g. Li_3), 12 (e.g. C_3 assuming that the lowest state from $(e')^2$
is of symmetry E'), 15 (e.g. N_3) and 21 (e.g. F_3) electrons will have a
Jahn-Teller cusp for D_{3h} configurations, whereas those with 6 (e.g. Be_3),
12 (e.g. B_3) and 18 (e.g. O_3) electrons would not. That does not mean
however that D_{3h} structures would be transition states for the exchange
reaction $X + X_2$ as we shall now see.

The Murrell-Laidler theorem [1] that the transition state force
constant matrix shall have only one negative eigenvalue stimulated
by a problem in transition state theory which was on the use of symmetry
numbers in calculating the partition function of the transition state.

To take a simple example [7], the rate constants for

$$H + H_2 \rightarrow H_2 + H \qquad\qquad\qquad\qquad i$$

and

$$H + D_2 \rightarrow HD + D \qquad\qquad\qquad\qquad ii$$

must be approximately the same, apart from the difference arising from
zero point vibrations and tunnel effects, but if in the transition state
expression for the rate constant

$$k = \frac{kT}{h} \frac{Q^{\neq}}{Q_A Q_B} \exp(-\varepsilon/kT)$$

the partition functions for the transition state $(Q^{\neq})$ and the reactants
$(Q_A Q_B)$ are calculated with the inclusion of the rotational symmetry
numbers, then the two rates differ by a factor of 2.

Schlag [8,9] and Bishop and Laidler [10] suggested a way around the
difficulties which was to omit all symmetry numbers but to multiply the
rate expression by a statistical factor which is the number of equivalent
activated complexes that can be found if all identical atoms in the
reactants are labelled. This recipe gave equal rate constants for
reactions i and ii because for both the reactants are of the type H + XY
and there are two equivalent transition states, namely HXY and HYX. It
also gives (correctly) half the rate for the reaction

$$H + HD \rightarrow H_2 + D \qquad\qquad\qquad\qquad iii$$

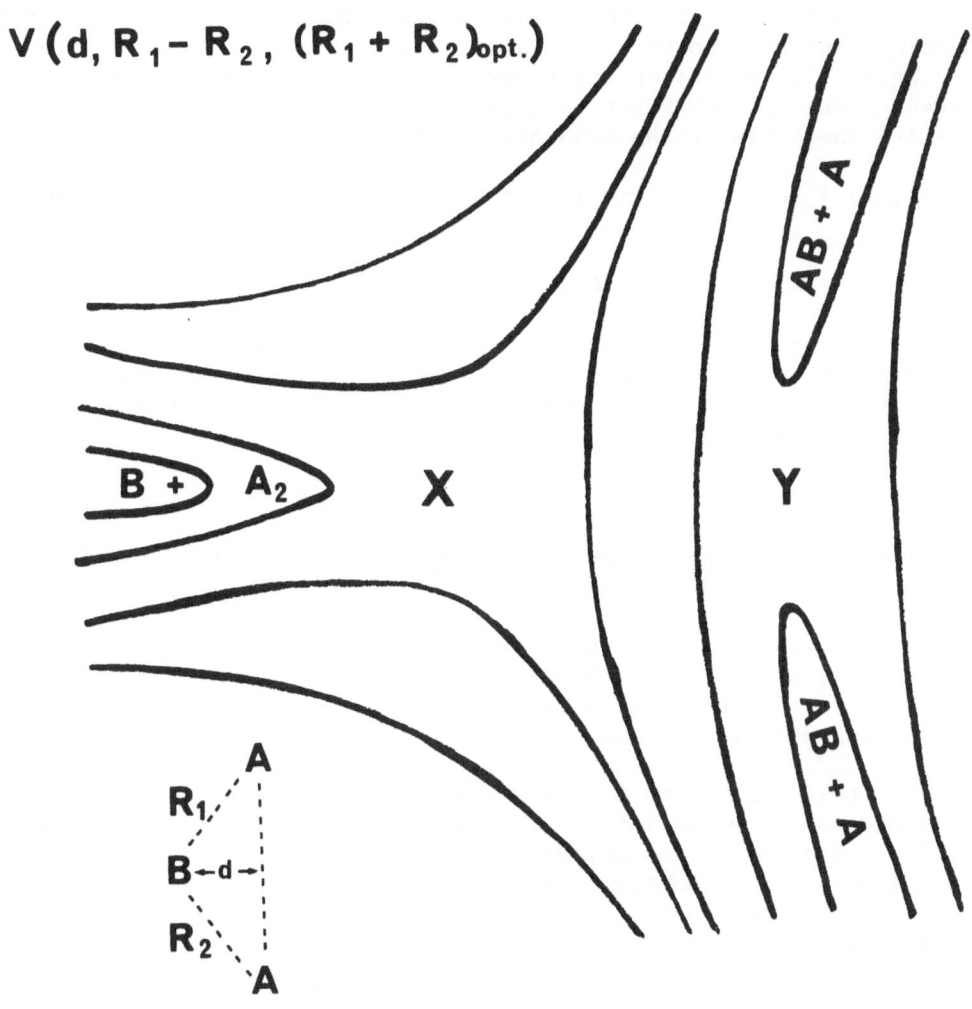

$$V(d, R_1 - R_2, (R_1 + R_2)_{opt.})$$

Figure 4. Contours of a schematic surface for the reaction $B+A_2 \to AB+A$. The points X and Y are saddle points.

because in this case there is only one transition state, namely HHD.

However, the Schlag-Bishop-Laidler recipe appeared to fail in some cases, because the ratio of rate constants for a forward and backward reaction did not equal the equilibrium constant between reactants and products as deduced from the ratio of partition functions with symmetry numbers. Murrell and Laidler suggested that in these cases of failure were because a transition state having too high a symmetry had been assumed. Murrell and Pratt [11] modified this conclusion for situations in which there was valley branching before or after a transition state.

Figure 4 is a schematic representation of the surface for the reaction

$$B + A_2 \rightarrow [A_2B] \rightarrow AB + A \qquad\qquad iv$$

where A_2B is a transition state having sufficiently high symmetry that both atoms A are equivalent. There are two saddle points for this reaction (labelled X and Y) one of which (Y) is the transition state for the exchange reaction

$$A + BA \rightarrow AB + A \qquad\qquad v$$

From the Murrell-Laidler theorem it can be deduced that the points X and Y do not coincide.

Murrell and Pratt showed that whether symmetry numbers or statistical factors give the correct rate expression from the reverse reaction to iv depends on whether the exchange reaction v is rapid compared with iv (as it generally is) or not.

The above discussion has been given not only to indicate the motiv- ation behind the Murrell-Laidler theorem but also to show that one must know the topology of the potential energy surface before transition state theory can be correctly applied.

A straightforward application of the Murrell-Laidler theorem is illustrated by figure 2. The question is whether the transition state for the isomerization of ozone

$$O_a - O_b - O_c \rightleftharpoons O_b - O_a - O_c \rightleftharpoons O_b - O_c - O_a$$

has D_{3h} symmetry. The normal modes of a D_{3h} structure are of type A_1' and E' as illustrated in figure 5. It is clear that the A_1' mode cannot be the reaction coordinate for isomerization as this leads to dissociation into three atoms. The E' mode will lower the symmetry from D_{3h} to C_{2v} but as this is doubly degenerate it cannot be a reaction coordinate or there would be two orthogonal directions of negative curvature. We therefore conclude that a D_{3h} structure can only be a metastable minimum on the surface, as has been suggested by some calculations [12-14] or a hill with two directions of negative curvature and the third curvature

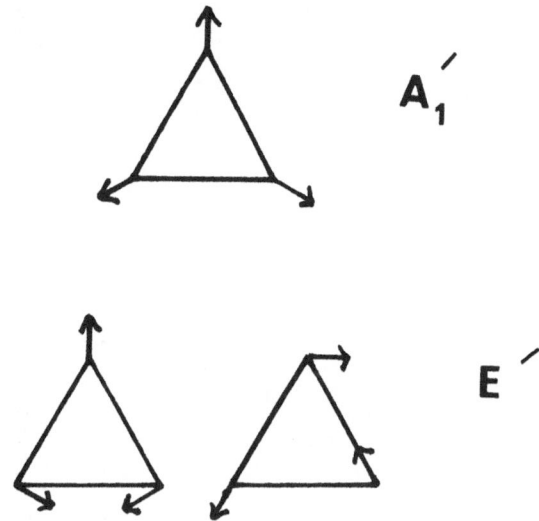

<u>Figure 5.</u>   Normal modes for an A_3 (D_{3h}) structure.

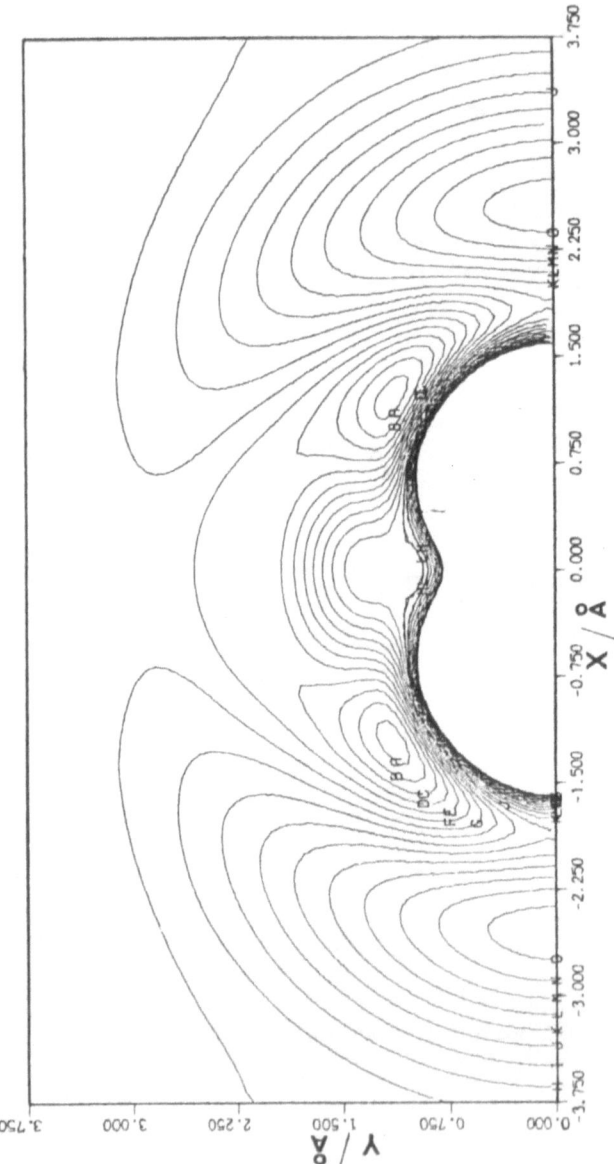

<u>Figure 6.</u> Contours of the potential energy surface for O_3 showing the
route for isomerization. The contours show the energy for one atom moving
around an O_2 molecule lying along the X axis, with centre of mass at X=0,
whose bond length is optimized for each position of the third atom.
Contour A= -6.25eV and successive contours are spaced by 0.25eV [3].

either positive or negative. The actual transition state for the ozone
isomerization is C_{2v} as shown in figure 6.

Stanton and McIver [15,16] formalized the Murrell-Laidler theorem
in three symmetry rules, the first of which we have already used. They
used the term <u>transition vector</u> for the normal coordinate of the transition state which has a negative eigenvalue (i.e. is parallel to the
reaction coordinate at the transition state).

1. The transition vector cannot belong to a degenerate representation
    of the group.

2. The transition vector is antisymmetric to any symmetry operation of
    the potential energy surface that converts reactants into products.

3. The transition vector is symmetric to any operation that leaves
    reactants or products unchanged, or changes reactants into equivalent
    reactants or products into equivalent products.

These theorems are perhaps all rather obvious but formal proofs have
been given by Stanton and McIver [15]. Rule 2 applies to what Salem
[17,18] has called a narcissistic reaction in which reactant and product
are related by a mirror plane or an S_n operation.

A more interesting theorem which is an extension of 2 and 3 above
concerns situations in which there is one or more equivalent reactant
or product. For example, suppose we have four equivalent valleys on a
surface which we identify with species X_1, X_2, X_3 and X_4. Let there be
a transition state T with a transition vector that is symmetric to
operations that convert $X_1 \rightarrow X_2$ and antisymmetric to operations that
convert $X_1 \rightarrow X_3$ or X_4. From the above rules 2 and 3 we know that T can
be a transition state for $X_1 \rightarrow X_3$ or X_4 but not for $X_1 \rightarrow X_2$. Moreover it
can be seen from figure 7 that in this situation there must be a lower
energy path for $X_1 \rightarrow X_2$ or $X_3 \rightarrow X_4$.

The above rules are illustrated by the following two examples; many
more have been given by Stanton and McIver [15].

1. <u>Inversion of tetrahedral carbon</u>

There has recently been considerable interest in the mechanism for
inversion of symmetry at a carbon atom and also in the possibility of
preparing a compound containing a tetravalent planar carbon atom [19].
It has been generally assumed that the transition state for inversion
of CH_4 would have D_{4h} symmetry. The out-of-plane bending modes of such
a structure are of symmetry A_{2u}, B_{1u} and E_g, and the B_{1u} mode would
be the transition vector. However, recent calculations [20] have
suggested that the A_{2u} mode also has a negative eigenvalue so that the
D_{4h} structure cannot be a transition state. Calculations on the surface
in the region of the D_{4h} structure has led to a transition state having
C_{4v} symmetry with HCH bond angles of about $80°$. The reaction would be

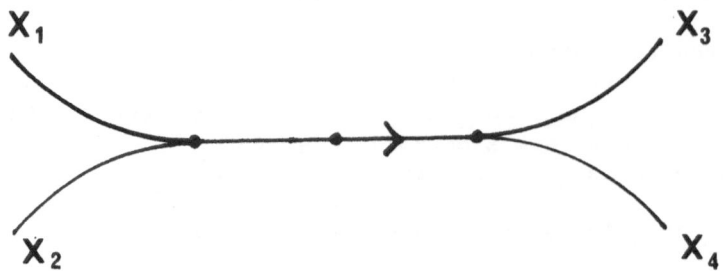

Figure 7. Schematic reaction path for a system showing symmetry equivalence of reactants (X_1, X_2) and products (X_3, X_4).

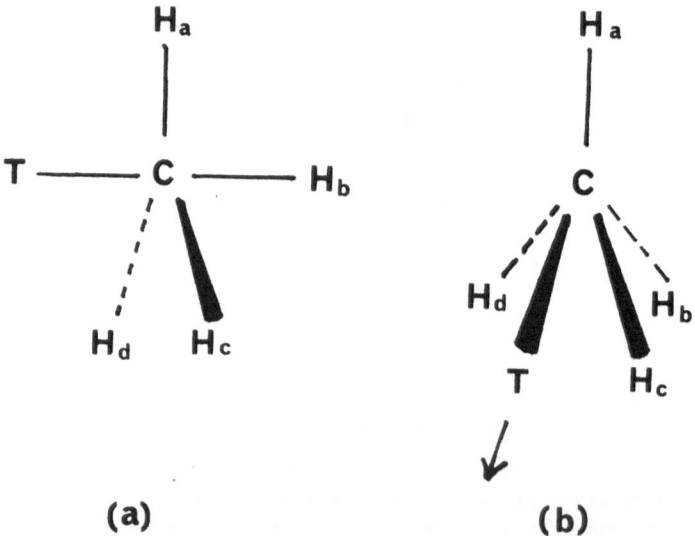

Figure 8. D_{3h} (a) and C_{4v} (b) structures which could be transition states for the S_{N2} substitution at a tetrahedral carbon atom.

non-synchronous in Salem's classification of narcissistic reactions [17,18]. Although no great reliability can be placed on such calculations they do illustrate the importance of calculating the normal coordinates of any proposed transition state.

2. Substitution at a tetrahedral carbon atom

S_{N2} substitution at a tetrahedral carbon atom is associated with inversion of configuration, and a D_{3h} (trigonal bipyramid) structure (fig. 8a) is consistent with this, the transition vector having A_1'' symmetry. However, the hot atom reaction

$$T + CH_4 \rightarrow CTH_3 + H$$

has been found (using substituted methanes) to take place with retention of configuration [20,21] and a C_{4v} transition state has been proposed for this [22]. The non-degenerate modes of such a structure are of A_1, B_1 and B_2 type (A_2 is a rotation about the 4-fold axis). The displacement arrow shown in figure 8b gives the reactant $T + CH_4$. Applying the C_{4v} symmetry operations to this arrow gives the following results [15]:

E and $\sigma_v (H_a CH_b)$ leave it unchanged

C_4 and σ_d (xz) $\rightarrow$ H_d + CH_4 with retention

C_4^{-1} and σ_d (yz) $\rightarrow$ H_c + CH_4 with retention

C_2 and σ_v $(H_c CH_d)$ $\rightarrow$ H_b + CH_4 with inversion

For the retention reaction the transition vector must be B_1 (-1 on C_4, -1 on σ_d, and +1 on σ_v. It cannot be a transition state for inversion because there is no non-degenerate irreducible representation which is antisymmetric under C_2. Moreover we conclude that if this structure is the transition state for the retention then there must be a lower energy transition state for the inversion of a different symmetry corresponding to the down-hill forking paths on each side of the C_{4v} structure. The D_{3h} structures would satisfy this conclusion.

3. Symmetry forbidden reactions and the pseudo Jahn-Teller effect

The application of orbital correlation diagrams to the elucidation of potential energy surfaces as in the derivation of the Woodward-Hoffmann rules for cycloaddition reactions [23] assumed that some elements of symmetry are retained along the reaction path. For a " symmetry forbidden" reaction there is then the question of whether the barrier may be removed by taking a path of lower symmetry. There are several studies to support this. One which we investigated was the reaction

$$H_2 + CH_2 \rightarrow CH_4$$

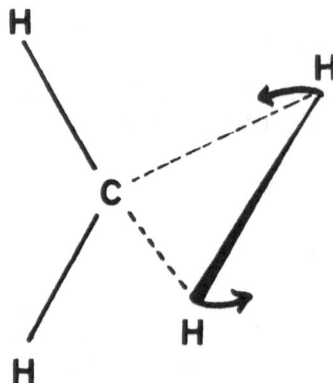

Figure 9. Preferred distortion from C_{2v} symmetry of the reaction path for $CH_2 + H_2 \rightarrow CH_4$ (singlet state) [24].

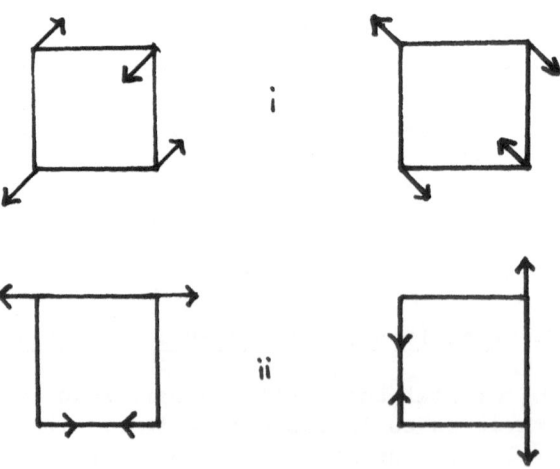

Figure 10. Two representations of the E_u modes of an A_4 (D_{4h}) molecule.

on both the singlet and triplet surfaces [24]. In both cases the
reaction was forbidden if the reaction path retained C_{2v} symmetry and
substantial barriers were obtained from ab-initio calculations along
this path. However, the barriers were almost completely removed by
allowing the structure to distort to a C_s symmetry as shown in figure 9
for the singlet state, and it was thought likely that it would be lost
completely if a further lowering of symmetry was allowed.

The distortion of a molecule to a lower symmetry structure in which
the change in energy is second-order in a nuclear displacement coordinate
is called the pseudo Jahn-Teller effect [25]. It is associated with the
second-order mixing of electronic states under the perturbation to the
Hamiltonian arising from the nuclear displacement with a stabilizion
which may or may not be larger than any first-order change in energy
associated with such a displacement.   There is no guarantee that the
overall effect of the displacement will be a stabilization.

Bader [26] was the first to recognize the role of the pseudo Jahn-
Teller effect for predicting reaction paths, and the idea has been developed by Salem [27,28], Bartell [29] and Pearson [30]. The pseudo Jahn-
Teller effect is most important for structures in which there are low-
lying excited states, and this is usually the situation for transition
states. To illustrate this we can consider the 4-centre transition state
for the $H_2 + D_2$ reaction.

There is a D_{4h} saddle point on the H_4 surface which could be the
lowest transition state for the $H_2 + D_2$ reaction. The question is whether
it is the lowest energy transition state. The electron configuration of
square H_4 is $a_{1g}^2 e_u^2$ which gives singlet states $^1A_{1g}$, $^1B_{1g}$ and $^1B_{2g}$.
The lowest of these is $^1B_{1g}$, and the two excited states are in order
$^1B_{2g}$ and $^1A_{1g}$ at the saddle point configuration [31]. The normal modes
of the D_{4h} structure are A_{1g}, B_{1g}, B_{2g} and E_g (in plane), and B_{1u} (out
of plane). There is no normal mode whose symmetry is the same as the
transition density of the first excited state ($B_{1g} \times B_{2g} = A_{2g}$) and the
second excited state ($B_{1g} \times A_{1g} = B_{1g}$) would mix with a B_{1g} mode which
is the symmetry of the transition vector. Thus neither of the first two
excited states will lead through a pseudo Jahn-Teller effect to a more
stable transition state.

The first excited configuration $a_{1g}^1 e_u^3$ gives a singlet state of E_u
symmetry and transitions to this will couple with the E_u in-plane
vibration. Two representations of the E_u mode are shown in figure 10.
One of these will lead to a kite structure and the other to a trapezium.
Both are allowed as transition states according to the Stanton-McIver
rules. Calculations [32] show that there is no saddle point on the kite
other than the D_{4h} structure but that the trapezoidal surface has a
saddle point considerably below the D_{4h} saddle. A reaction path for the
4-centre reaction starts at rectangular configurations, passes through
the trapezoidal saddle point and exits via a linear configuration of the
four atoms [4].

' If these are obtained by the standard FG matrix method [2], the normal coordinates will be combinations of mass weighted internal coordinates. However, the masses can always be subsumed in the coefficients multiplying the internal coordinates.

REFERENCES

1. Murrell,J.N., and Laidler,K.J., 1968, Trans.Faraday Soc., 64, pp.371.
2. Wilson,E.B., Decius.J.C., and Cross,P.C., 1955, Molecular Vibrations, McGraw-Hill.
3. Murrell,J.N., Sorbie,K.S., and Varandas,A.J.C., 1976, Molec.Phys. 32, pp.1359.
4. Varandas,A.J.C., and Murrell,J.N., 1977, Disc.Faraday Soc., 62, pp.92-109.
5. Jahn,H.A., and Teller,E., 1937, Proc.Roy.Soc., 161A, pp.220.
6. Murrell,J.N., 1972, J.C.S.Chem.Comm., pp.1044.
7. Benson,G.W., 1958, J.Amer.Chem.Soc., 80, pp.5151.
8. Schlag,E.W., 1963, J.Chem.Phys., 38, pp.2480.
9. Schlag,E.W., and Haller,G.L., 1965, 42, 584.
10. Bishop,D.M., and Laidler,K.J., 1965, J.Chem.Phys., 42, pp.1688.
11. Murrell,J.N., and Pratt,G.L., 1970, Trans.Faraday Soc., 66, pp.1680.
12. Wright,J.S., 1973, Can.J.Chem., 51, pp.139.
13. Shih,S., Buenker,R.J., and Peyerimhoff,S.D., 1974, Chem.Phys.Lett., 28, pp.463.
14. Hay,P.J., Dunning,T.H., and Goddard,W.A., 1975, J.Chem.Phys., 62, pp.3912.
15. Stanton,R.E., and McIver,J.W., 1975, J.Amer.Chem.Soc., 97, pp.3632.
16. McIver,J.W., 1974, Accts.Chem.Res., 7, pp.72.
17. Salem,L., 1971, Accts.Chem.Res., 4, pp.322.
18. Salem,L., Durup,J., Bergeron,C., Cazes,D., Chapuisat,X., and Kagan.H. 1972, J.Amer.Chem.Soc., 92, pp.4472.
19. Hoffmann,R., Alder,R.W., and Wilcox,C.F., 1970., J.Amer.Chem.Soc., 92, pp.4492.
20. Henchman,M., and Wolfgang,R., 1961, J.Amer.Chem.Soc., 83, pp.2991.
21. Palino,G.F., and Rowland,F.S., 1971, J.Phys.Chem., 75, pp.1299.
22. Weston,R.E., and Ehrenson,S., 1971, Chem.Phys.Lett., 9, pp.351.
23. Woodward,R.B., and Hoffman,R., 1965, J.Amer.Chem.Soc., 87, pp.2511.
24. Murrell,J.N., Pedley,J.B., and Durmaz,S., 1973, J.C.S.Faraday II, 69, pp.1370.
25. Öpik,U., and Pryce,M.H.L., 1957, Proc.Roy.Soc., A238, pp.425.
26. Bader,R.F.W., 1962, Can.J.Chem., 40, pp.1164.
27. Salem,L., 1969, Chem.Phys.Lett., 3, 99.
28. Salem,L., and Wright,J.S., 1969, J.Amer.Chem.Soc., 91, pp.5947.
29. Bartell,L.S., 1963, J.Chem.Ed., 45, pp.754.
30. Pearson,R.G., 1969, J.Amer.Chem.Soc., 91, pp.1252,4947.
31. Conroy,H. and Malli,G., 1969, J.Chem.Phys., 50, pp.5049.
32. Silver,D.M. and Stevens,R.M., 1973, J.Chem.Phys., 59, pp.3378.

ON THE USE OF AROMATICITY RULES, FRONTIER ORBITALS AND
CORRELATIONS DIAGRAMS. SOME DIFFICULTIES AND UNSOLVED
PROBLEMS.

NGUYEN TRONG ANH
Laboratoire de Chimie Théorique
Université de Paris-Sud   (490)
91405   ORSAY (France)

ABSTRACT

The most usual difficulties encountered in the study of frontier-controlled reactions by approximate methods (aromaticity rules, frontier-orbital approximation and correlation diagram) are reviewed.

x x x

Since the discovery of the Woodward-Hoffmann rules [1], three approximate theoretical methods [1-4] have been extensively used by chemists for the study of chemical reactivity. These methods are based on the aromaticity rules [2], on the frontier orbital approximation |3| , and on correlation diagrams [1]. The application of these methods, usually very simple, may sometimes presents serious difficulties. This paper is concerned with such cases.

1. AROMATICITY RULES

The theoretical justifications of aromaticity rules have been given only for annulenes, i.e. for monocyclic polymethine chains $(CH)_n$ where each carbon atom contributes only one atomic orbital to the conjugating system. It is then clear that the aromaticity rules may not be obeyed if one of the three starting assumptions is not satisfied. More precisely, one should not, without due caution, extend the aromaticity rules to
- polycyclic systems
- heteroatomic systems (all conjugated atoms should be identical)
- systems in which some atom(s) may use more than one atomic orbital in the reaction (cumulenes, acetylenes, transition metal complexes)
- cheletropic and related processes where extracyclic atoms undergo important changes during the reaction

Let us consider the 6+4 thermal cycloaddition of fulvene with butadiene. Normally, as the transition state 1 is bicyclic, aromaticity rules cannot be used to predict the stereochemistry of this reaction. An ingenious way to by-pass this difficulty has been given by Dewar [6]. One first considers the (10) annulene 2. This being an alternant hydro-

177

R. Daudel, A. Pullman, L. Salem, and A. Veillard (eds.), Quantum Theory of Chemical Reactions, Volume I, 177-189.
Copyright © 1979 by D. Reidel Publishing Company.

carbon, the bond order between 2 and 6, both non-starred atoms, is zero.
Therefore, azulene 3 has the same π stabilization energy as 2 and should
be aromatic in a Hückel topology. The stereochemistry of the cycloaddition
(fulvene + butadiene) should then be supra-supra. However, we are unable,
by this method, to predict whether fulvene would prefer an electron-rich

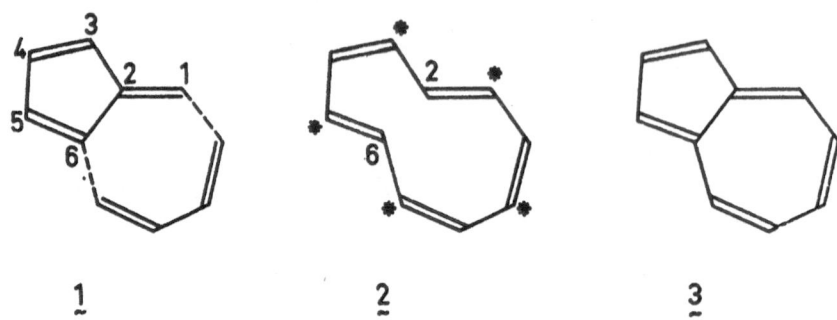

or an electron-poor diene. Now the fulvene HOMO 4 shows a zero coefficient at atom 1 which is a reacting center in the 6+4 cycloaddition.
Therefore, the rate should increase if fulvene intervenes essentially
by its LUMO 5, i.e. if the second addend is an electron rich diene.
Indeed, Houk [7] has found that fulvene reacts as a 6e component with
electron-rich dienes (1-aminobutadiene) and as a 2e component with
"normal" or electron-poor dienes (cyclopentadiene, cyclopentadienones,
α-pyrones...).

The previous example nicely illustrates two points. First, it is
not unfrequent that the frontier orbital approximation gives more precise
informations than the aromaticity rules. The loss of finer details is the
price we have to pay sometimes for the simplicity of the latter approach.
Second, it is possible to extend the Dewar-Zimmerman treatment to some
polycyclic transition states. However, this extension is not always
straightforward. An example is the bicyclobutane-butadiene interconversion.
If we look at the ring opening reaction (6 → 7), we arrive without difficulty to the correct conclusions. The transition state is assimilated to

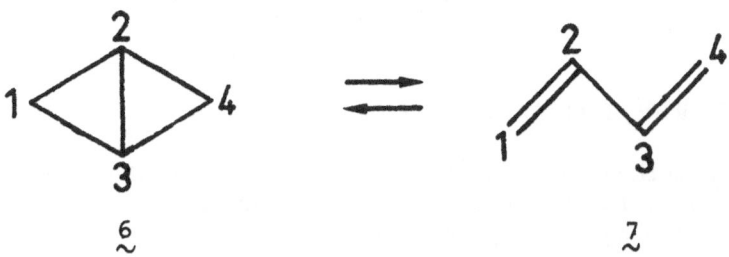

a cyclobutadiene derivative in which a bridge has been created between two opposite apices. It has therefore the same π stabilization as cyclobutadiene itself and the thermal reaction will be allowed if the orbital array has a Möbius topology. Consider now the reverse reaction (7 → 6). Butadiene being alternant, the bond orders between the two pairs of atoms (1 and 3) and (2 and 4) are zero. It is then tempting to conclude that the transition state resembles a non-aromatic annulene and that the reaction will be non stereoselective, which is in contradiction with experimental results [8].

Cases are known in which an inconsiderate use of aromaticity rules may lead to completely erroneous conclusions. One example is the hydroboration of olefin which corresponds formally to a syn-addition of boron and hydrogen to the double bond. "All the available evidence suggests that it is a concerted process and takes place through a cyclic four-membered transition state" [9]. Therefore, it has sometimes been suggested that this reaction violates the Woodward-Hoffmann rules ; its transition state being similar to a Hückel cyclobutadiene. This argument is incorrect : the MO's of a cyclic transition state do <u>not</u> resemble the MO's of an annulene if <u>one atom (or more) of the cycle uses more than one atomic orbital</u> [5-10]. Consider the hydroboration transition state 8 : it contains four atoms, four electrons and <u>five</u> atomic orbitals. The overlap between the orthogonal orbitals 1 and 5 is zero. Thus the MO's of this cyclic transition state resemble in fact the MO's of the

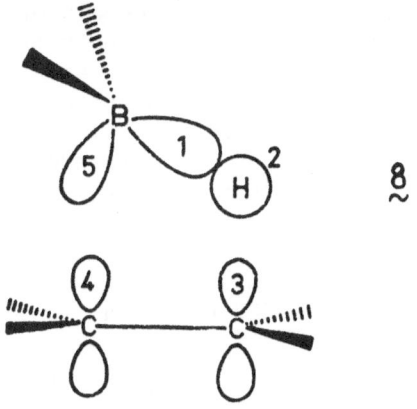

open-chain pentadienyl cation ! And clearly it is meaningless to talk
about the aromaticity or antiaromaticity of an open-chain polyene. A
similar reasoning shows that the concerted unimolecular elimination of
hydrogen chloride from ethylchloride is not a pericyclic reaction [11].

2. FRONTIER ORBITAL APPROXIMATION

This method has been much criticized. One of the major objections
is that molecular orbitals being mathematical functions without precise
physical meaning, it may be dangerous to select a few orbitals and to
reason on them as if they were observables. Let us examine more closely
this objection.

For Hückel type calculations, the total energy is equal to the
sum of orbital energies. If in this sum there are some predominant terms,
they suffice to give the general trend and the neglect of other terms
is justified. According to a model calculation by Herndon and Hall [12],
in Diels-Alder reactions, the frontier orbitals account for 64 % up to
82 % of the total energy change. The frontier orbital theory is in this
case a fully legitimate approximation.

The situation is less clear with Hartree-Fock SCF calculations
where the total energy is no longer equal to the sum of orbital energies.
Instead :

$$E = \sum_{k} n_k \, \varepsilon_k - V_{ee} + V_{nn}$$

where the ε_k's are the canonical orbital energies, n_k the orbital occu-
pation numbers, V_{ee} the electron-electron and V_{nn} the nuclear-nuclear
repulsion energies. It is not obvious why the variation of the total
energy should follow closely the variations of the energies of a few
orbitals. It may be noted that this difficulty is not specific to frontier orbital theory but is also encountered in the construction of
Mulliken Walsh diagrams : when using Hartree-Fock SCF calculations,
what is to be identified as the ordinate in these diagrams ? This problem
has received the attention of many theoreticians [13] and several ingenious energy partitioning schemes have been proposed, from which we may
single out those of Davidson [14] and of Hoffmann [15]. Davidson has
defined a set of internally consistent SCF energies the sum of which
is exactly equal to the total energy. Hoffmann defined an "average
state" of a molecule by distributing the electrons equally among the
valence orbitals of a minimal basis set Hartree-Fock calculations. The
resulting eigenvalues, called "tempered orbital energies", behave much
more like the Mulliken-Walsh diagram energies or EH eigenvalues than
do the HF canonical orbital energies. Interestingly, in the few cases
studied so far, the occupied tempered orbital energies are not very
different from the respective SCF canonical orbital energies.

Buenker et al. [16] have examined the possibility that the energy
change of some "critical orbitals" may reflect the total energy change.

At least for the cyclobutene-butadiene thermal interconversions, very good agreement was found to exist between total and critical orbital energies for the disrotatory and the conrotatory modes.

Still, several serious problems remain. First, can we extend the frontier orbital approximation to reactions of anions ? This is an important practical question, as ionic reactions constitute a major part of Organic Chemistry. Hückel type calculations do not take into account bielectronic repulsion terms and give therefore rather unsatisfactory representations of anions. In particular they cannot reproduce the reversal of the nucleophilicity order when one goes from a protic solution to a dipolar aprotic solution [17]. Hartree-Fock SCF calculations give rise to two types of difficulties. When small basis sets are used, the highest occupied orbitals of anions often have positive energies, which are meaningless. Furthermore, the level ordering may change with the basis set. For example, the HOAO (highest occupied atomic orbital) of F^- is at higher energy than the HOAO of Cl^- if minimal basis sets are used, and is at lower energy if extended basis sets are used [17-18].

It may be recalled here that the interaction of a HOMO ϕ_A with a LUMO ϕ_B leads to a charge tranfer from A to B : the two electrons, initially localized in A, occupy after interaction the orbital $\phi_A^1 = \phi_A + \lambda\phi_B$ and are therefore partially delocalized into B (Figure I).

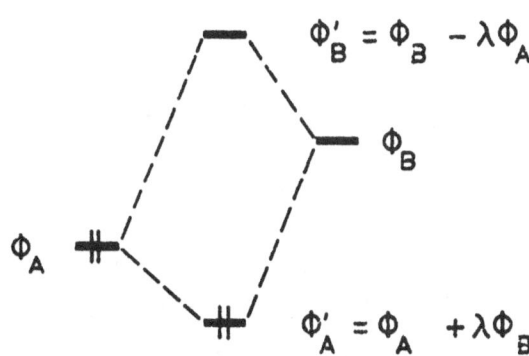

$$\phi_B' = \phi_B - \lambda\phi_A$$
$$\phi_B$$
$$\phi_A$$
$$\phi_A' = \phi_A + \lambda\phi_B$$

Figure I - Interaction of an occupied orbital and a vacant orbital.

For a given ϕ_B, the higher ϕ_A, the larger the charge transfer, provided that the overlap remains the same. This means that, in the frontier orbital approximation, the HOMO level is a measure of the donor character of A. Now it seems reasonable to consider that ionization potentials also measure the donor character of molecules. It follows that, for internal consistency, one should use for orbital interactions only the molecular orbitals which satisfy qualitatively the Koopman's theorem, i.e. the compound with the higher HOMO should have the lower first ionization potential [19-20].

Before we leave this subject, mention should be made of the now well-known cases of subjacent and superjacent orbital controls [22], where the frontier orbital approximation fails. A typical example of superjacent orbital control is the cycloaddition of ketene. This compound reacts preferentially with electron-rich olefins to give cyclobutanones 9 and not α-methylene oxetane 10 as would be expected from a brutal application of frontier orbital approximation (Figure II).

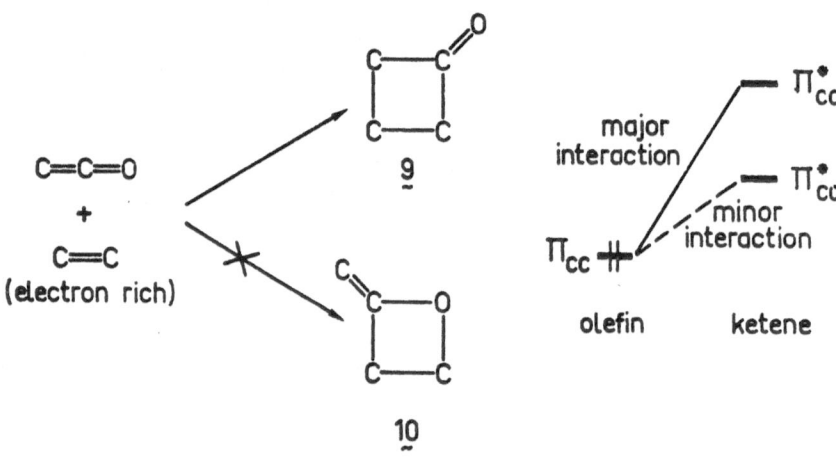

Figure II – Superjacent orbital control in ketene cycloaddition. The major interaction occurs between the olefin HOMO and the ketene π^*_{CC}, and not with the ketene LUMO π^*_{CO}.

3. CORRELATION DIAGRAMS

Only orbital correlation diagrams are discussed in this section but the remarks made here apply to state correlation diagrams as well.

The usual way to construct correlation diagrams is to classify the MO of the reagents and of the products according to their energies and their symmetries. The correlation lines are then drawn, which connect MO of like symmetry. When there are more than one MO of a given symmetry, the indetermination is lifted with the help of the non-crossing rule : two correlation lines of like symmetry should not cross [1]. The thermal reaction is considered to be symmetry-allowed if one occupied orbital always correlates with another occupied orbital. If there is one (occupied-vacant) correlation, the thermal reaction is forbidden and the photochemical reaction symmetry allowed.

All the difficulty rests on the determination of the reaction symmetry elements. Usually a reaction has less symmetry than the initial
and final products. For example, cis-butadiene and cyclobutene belong to
the C_{2v} group, but the transition state for their electrocyclic transformation belongs either to the C_s (disrotatory mode) or to the C_2 group
(conrotatory mode). However, for intermolecular reactions or for reactions with non rigid molecules, the transition state may be the species
with the highest symmetry. This is not really a problem. As pointed out
by Hsu et al.[23], correlation diagrams should be drawn for reactants and
products in their activated structures, not in their ground state geometries. One may also use non-rigid molecule group theory [24]. A more
serious problem is raised by reactions in which the symmetry of the transition state must be artifically increased. Consider the cycloaddition of
a carbonyl with a CC double bond to give oxetane 11. It is well known that

$$ \underset{\underset{\sim}{11}}{} $$

this reaction occurs photochemically but not thermally [25]. If a correlation diagram is constructed using the only symmetry element of the
system, i.e. the plane containing the 4 heavy atoms, it is found that the
reaction is thermally allowed. To get a better agreement with experience,
it is necessary to replace the oxygen atom by a methylene group, i.e; to
treat this reaction as if it were an ethylene cyclodimerization.

The trouble is that in other cases, the symmetry of the system must
be reduced. Consider the one step formation of cubane 13 from cycloocta-
tetraene 12. A correlation diagram, which takes into account all the

12   13

symmetry elements of the cube, leads to the conclusion that the reaction
is thermally allowed. Woodward and Hoffmann [27] pointed out that this
reaction should in fact be considered as a superposition of two 2+2
cycloadditions [28].

   The worse is yet to come ! Let us return to the $6 \rightleftarrows 7$ interconversion [30]. The only molecular symmetry element is a twofold axis
passing through the middle of the C_2-C_3 single bond. A correlation diagram using this axis would give incorrect predictions. To reproduce the
experimental results, one should discard this real molecular symmetry
axis, consider instead a straight line passing through the middles of
segments 1-2 and 3-4 (or 1-3 and 2-4) and distort the molecule to make
this an axis of symmetry. Needless to say, these acrobatic technics in
the hand of beginners may lead to disastrous results.

   A similar but more complex case is the benzvalene-benzene isomerization ($14 \rightarrow 15$). Again, two correlation diagrams are possible [31].
The one using the molecular symmetry axis bisecting bonds 2-3 and 5-6
predicts an allowed reaction. The one using the artificial symmetry
axis bisecting bonds 1-3 and 2-4 indicates a forbidden reaction. As a
matter of fact, all depends on the role of the 5-6 double bond in the
transition state. If the transition state occurs at an early stage,
this role is negligible and it is clear that this reaction is very

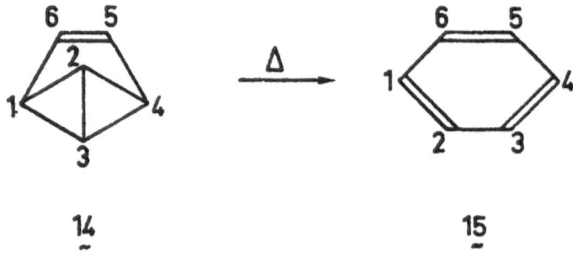

$$14 \qquad\qquad\qquad\qquad 15$$

similar to the bicyclobutane-butadiene isomerization and will be forbidden. If however the transition state is a late one, a stronger inter-
vention of the 5-6 double bond may be expected and the reaction will be
allowed [32]. In other words, knowledge of the reaction symmetry elements
is here insufficient and more information is needed to conclude. For
further discussion, see [33].

   The difficulties encountered in the construction of correlation
diagrams and the precautions which should hence be taken have been
extensively discussed by Woodward and Hoffmann [34]. They pointed out
that "the symmetry elements chosen for analysis must bisect bonds made
or broken in the process". Applying this rule, we can indeed study
without much difficulty all the foregoing examples as well as many
other reactions (all pericyclic and cheletropic reactions in fact). It
should be noted however that the correlation diagrams which have been
used successfully by Salem for the study of photochemical reactions [35]

and mass spectrometric fragmentation [36] do not use symmetry elements which bisect bonds made or broken in the process. The reason is that in these cases, the most important change concerns the number of π and of σ electrons and the molecular plane is the symmetry element which best discriminates them. Clearly the construction of correlation diagrams has not reached the stage where one could apply mechanically some well-defined rules. Before the diagram is constructed, the reaction should be carefully analyzed from the chemist's point of view to determine what are the most important changes. These changes in turn will determine the choice of the symmetry elements.

As a final example, let us analyze Mango's theory of metathesis [37]. An essential feature of this theory is that the cyclodimerization of ethylene, which is thermally forbidden (a bonding combination of the ethylenes MO, of AS symmetry, correlating with an antibonding MO of cyclobutane) becomes symmetry-allowed in the presence of a transition metal atom. If the catalyst possesses a vacant d orbital of AS symmetry and an occupied d orbital of SA symmetry, the bonding-antibonding correlations are avoided and replaced by bonding - non bonding correlations (Figure III). Unfortunately, this attractive theory is contradicted by experimental tests [38] and must be discarded. The main problem however remains : the correlation diagram shown in Figure III looks like a bona fide Woodward-Hoffmann diagram, and nevertheless, gives incorrect predictions. Therefore, either there is some hidden mistake in this diagram or all Woodward-Hoffmann diagrams are questionable.

Considering the successes of the Woodward-Hoffmann rules, it seems logical to explore first the former hypothesis. Many possible arguments may be advances, one of the most reasonable derives from the fact that no transition metal complex of cyclobutane has ever been observed. Figure III should then be replaced by Figure IV. The crucial cyclisation step requires little energy according to the diagram of Figure III : the cost in energy necessary for the transformation of a bonding AS orbital into a non-bonding one is compensated by the gain in energy coming from the SA non-bonding/bonding correlation. In Figure IV, both occupied SA orbitals are bonding ones and there is no full compensation. If further works confirm this interpretation, we have here another example where the erroneous conclusions proceed from an unrealistic chemical model, the construction of the correlation diagram and its interpretation being perfectly correct.

4. ACKNOWLEDGEMENTS

The author is indebted to Professor J. Michl and to Dr. G. Berthier for a discussion on the Koopman's theorem and on the Hartree-Fock instability problem, and to Professor R. Hoffmann for a preprint of his paper.

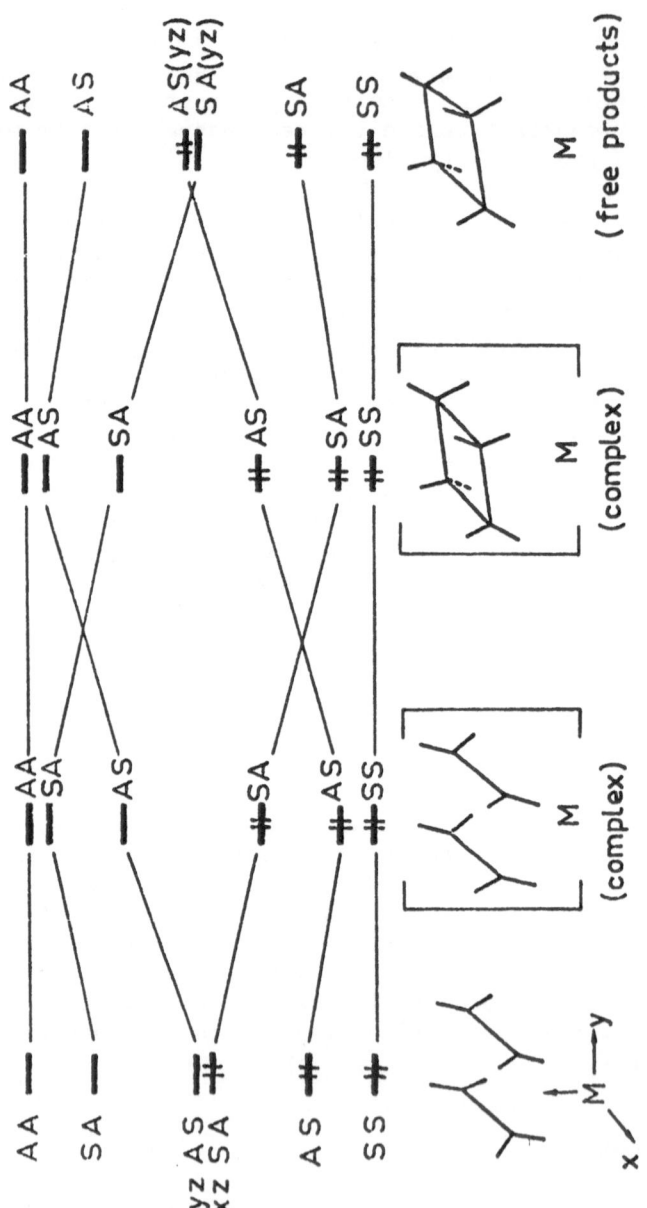

Figure III – Correlation diagram for olefin metathesis according to Mango's theory.

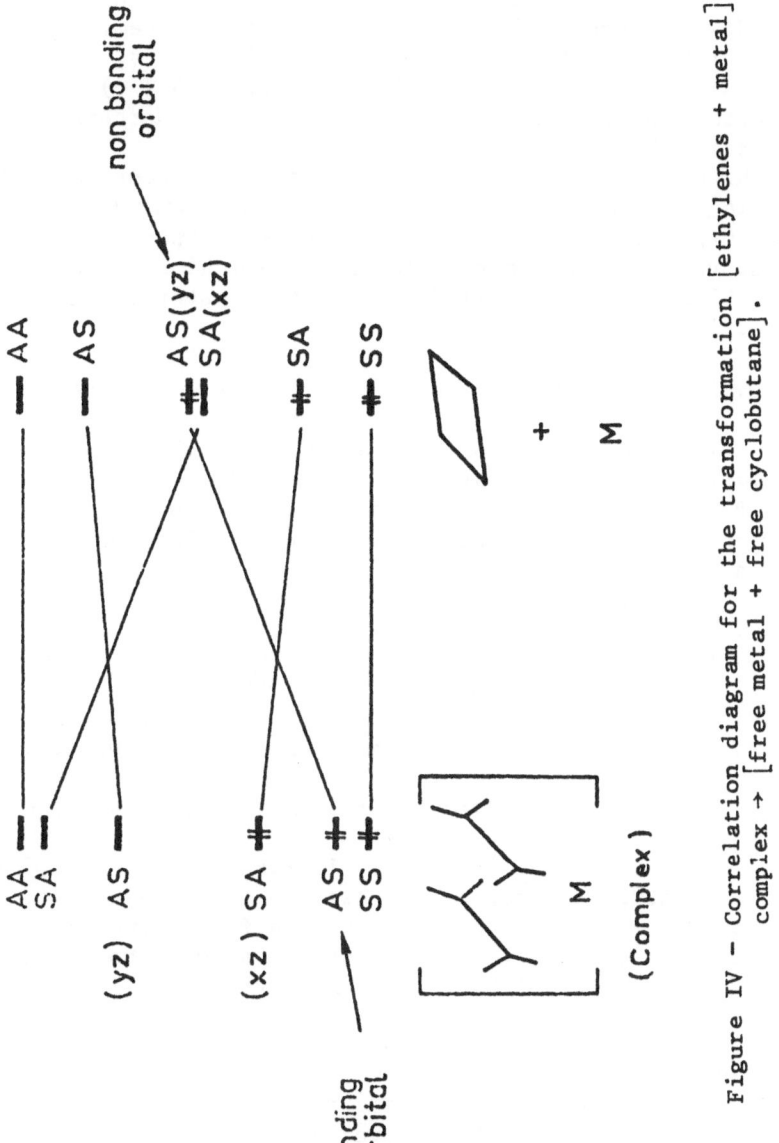

Figure IV - Correlation diagram for the transformation [ethylenes + metal complex → [free metal + free cyclobutane].

REFERENCES

1. a) R.B. Woodward and R. Hoffmann, "The conservation of Orbital Symmetry", Verlag Chemie, Academic Press, Weinheim 1970 ; b) Nguyên Trong Anh, "Les Règles de Woodward-Hoffmann", Ediscience, Paris 1970
2. a) M.J.S. Dewar, "The MO Theory of Organic Chemistry", McGraw-Hill, N.Y. 1969 ; b) M.J.S. Dewar and R.C. Dougherty, "The PMO Theory of Organic Chemistry", Plenum Press, N.Y. 1975 ; c) H.E. Zimmerman, in "Pericyclic Reactions", R.E. Lehr and A.P. Marchand Eds., Academic Press, N.Y. 1977, Vol. I, p. 53
3. a) K. Fukui, "Theory of Orientation and Stereoselection", Springer-Verlag, Berlin, Heidelberg 1975 ; b) I. Fleming, "Frontier Orbitals and Organic Chemical Reactions", Wiley, London 1976
4. R.G. Pearson, "Symmetry Rules for Chemical Reactions. Orbital Topology and Elementary Processes", Wiley, Chichester 1976
5. Ref. 1b, pp. 20, 146 and 156
6. Ref. 2a, p. 222
7. K.N. Houk, J.K. George and R.E. Duke, Jr., Tetrahedron, $\underline{30}$, 523 (1974)
8. G.L. Closs and P.E. Pfeffer, J. Amer. Chem. Soc. $\underline{90}$, 2452 (1968)
9. W. Carruthers, "Some modern methods of organic synthesis", Cambridge University Press, Cambridge 1971, p. 214
10. a) R.A. Jackson, J. Chem. Soc. (B) 58 (1970) ; b) E.A. Hill, J. Organomet. Chem. $\underline{91}$, 222 (1975) ; c) T. Clark and P.v.R. Schleyer, ibid. $\underline{156}$, 191 (1978)
11. P.C. Hiberty, J. Amer. Chem. Soc. $\underline{97}$, 5975 (1975)
12. W.C. Herndon and L.H. Hall, Theor. Chim. Acta, $\underline{7}$, 4 (1967)
13. R.J. Buenker and S.D. Peyerimhoff, Chem. Rev. $\underline{74}$, 127 (1974)
14. E.R. Davidson, J. Chem. Phys. $\underline{57}$, 1999 (1972) ; L.Z. Stenkamp and E.R. Davidson, Theor. Chim. Acta $\underline{30}$, 283 (1973) ; E.R. Davidson and L.Z. Stenkamp; Intern. J. Quant. Chem. Symp. $\underline{10}$, 21 (1976)
15. P.K. Mehrotra and R. Hoffmann, Theor. Chim. Acta $\underline{48}$, 301 (1978)
16. R.J. Buenker, S.D. Peyerimhoff and K. Hsu, J. Amer. Chem. Soc. $\underline{93}$, 5005 (1971)
17. C. Minot and Nguyên Trong Anh, Tetrahedron Lett. 3905 (1975)
18. E. Clementi, IBM Journal of Research and Development, $\underline{9}$, 2 (1965)
19. In particular, the orbitals obtained for F^- and Cl^- with extended basis set [18] should not be used in a perturbational scheme, the electron affinities for F and Cl being respectively 3.48 eV and 3.69 eV [21]
20. In connection with this problem, it may be noted that if in the gas phase or in aprotic solutions, the HOAO of F^- is higher than that of Cl^-, the order is reversed in protic solutions. Similarly there is a reversal of the LUAO level ordering for the Li^+, Na^+ pair, when the solvent is changed [17]. Therefore, the equivalences suggested by Klopman [21]
        hard cation + hard anion = charge controlled reaction,
         soft cation + soft anion = frontier controlled reaction,
   hold only for protic solutions or for reactions with aggregates [17]
21. G. Klopman, J. Amer. Chem. Soc. $\underline{90}$, 223 (1968)
22. J.A. Berson and L. Salem, J. Amer. Chem. Soc. $\underline{94}$, 8917 (1972) ;

W.T. Borden and L. Salem, ibid. 95, 932 (1973) ; S. David,
O. Eisenstein, W.J. Hehre,L. Salem and R. Hoffmann, ibid. 95, 3806
(1973) ; J.A. Berson, Accts. Chem. Res. 5, 405 (1972) ; L. Salem,
"Chemical and Biochemical Reactivity", The Jerusalem Sympsoia on
Quantum Chemistry and Biochemistry VI, The Isreal Academy of
Sciences and Humanities, Jerusalem 1974, p. 329

23. K. Hsu, R.J. Buenker and S.D. Peyerimhoff, J. Amer. Chem. Soc. 93,
    2117 (1971)

24. For an example, see : Y. Ellinger and J. Serre, Int. J. Quant. Chem.
    7S, 217 (1973)

25. The reactive state for the Paterno-Büchi reaction is most probably
    a n, π^* triplet and not a π, π^* singlet [26]. However, the main
    point of our argument is that experimentally the thermal reaction
    does not occur

26. D.R. Arnolds, "Adv. in Photochemistry", W.A. Noyes, Jr., G.S.Hammond,
    J.N. Pitts, Jr., Eds. Interscience, N.Y. 1968, Vol. 6, p. 301

27. Ref. 1a, p. 32. See also : J. Langlet and J.P. Malrieu, J. Amer. Chem.
    Soc. 94, 7254 (1972)

28. Note also that in intermolecular reactions, for large internuclear
    distances, molecular orbitals with reduced (broken) symmetry may
    give a lower total energy than symmetry-adapted MO's |29|. Again
    this difficulty may be avoided by considering the reagents in their
    activated forms

29. J.M. McKelvey and G. Berthier, Chem. Phys. Lett. 41, 476 (1976) ;
    J. Paldus and A. Veillard, ibid. 50, 6 (1977)

30. Ref. 1a, pp. 34, 76-78

31. E.A. Halevi, Nouv. J. Chim. 1, 229 (1977)

32. It is easy to see that the reaction then resembles a (4a+2a) cyclo-
    addition with the 2-3 fragment playing the role of the 2e component

33. J.J.C. Mulder, J. Amer. Chem. Soc. 99, 5177 (1977)

34. Ref. 1a, p. 31

35. L. Salem, J. Amer. Chem. Soc. 96, 3486 (1974) ; Israel J. Chem. 14,
    (1975) ; Science 191, 822 (1976) .; W.G. Dauben, L. Salem and
    N.J. Turro, Accts. Chem. Res. 8, 41 (1975)

36. C. Minot, Nguyên Trong Anh and L. Salem, J. Amer. Chem. Soc. 98,
    2678 (1976)

37. F.D. Mango and J.H. Schachtschneider, J. Amer. Chem. Soc. 89, 2484
    (1967)

38. See, for an example : P. Heimbach, Angew. Chem. Intern. Ed. 12, 975
    (1973)

ON THE USE OF THE ELECTROSTATIC MOLECULAR POTENTIAL IN THEORETICAL
INVESTIGATIONS ON CHEMICAL REACTIVITY

JACOPO TOMASI
*Laboratorio di Chimica Quantistica ed Energetica Molecolare
del C.N.R., Via Risorgimento 35, 56100 Pisa, Italy.*

ABSTRACT. The validity and the limits of the methods which use the electrostatic molecular potential for the study of chemical reactivity are
discussed. The logical connection of these methods with the complete
calculation of the reactive interaction energy and to the large family
of reactivity indexes is analyzed. A tentative classification of reactions in a few groups where the electrostatic approximation may play a
different role is presented, and a review is given of the most outstanding results thus far obtained, with special emphasis to hydrogen bonding, interactions with an atomic cation, S_E2 and S_N2 reactions, and photochemical reactions. Finally practical methods which greatly reduce
the computation time with respect to more straightforward applications
of the electrostatic approximation are presented.

1. INTRODUCTION

The number and variety of approximate methods which introduce electrostatic considerations in a theory of chemical reactivity is so large
that one cannot give, within the limits imposed by a conference, a critical and sufficiently exhaustive resumé of this topic.
    I shall limit my exposition to the more specific subject suggested
by the title, i.e. the examination of methods for the study of chemical
reactivity which make explicit use of the electrostatic molecular potential V, calculated as one-electron observable from a given molecular
wavefunction. I shall also exclude any discussion on the usefulness of
this quantity in other fields of research, such as the interpretation
of electron-molecule scattering phenomena |1|, the determination of the
properties of molecular crystals |2|, the study of environmental effects
due to solvation |3|, etc.

2. THE CONNECTION BETWEEN ELECTROSTATIC POTENTIAL AND REACTIVE INTERAC-
    TION ENERGY

The most direct way of introducing V in studies of chemical reactivity

R. Daudel, A. Pullman, L. Salem, and A. Veillard (eds.), Quantum Theory of Chemical Reactions, Volume I, 191-228.
Copyright © 1979 by D. Reidel Publishing Company.

is by way of a consideration of the well known and widely employed decomposition of the interaction energy ΔE_{AB} between two reacting molecules |4,5|:

$$\Delta E_{AB} = E_{AB} - (E_A^\circ + E_B^\circ) = E_{es} + E_{pol} + E_{ct} + E_{exch} + E_{disp} \tag{1}$$

Eq.(1) corresponds to an operationally simple manner of deducing from ab initio computations on the supermolecule AB (with internal geometries of the two subunits A and B kept fixed at those selected for infinite A-B separation) a decomposition of ΔE into physically meaningful terms. A schematic outline of the decomposition procedure in the SCF framework is given in Table I (of course in the SCF scheme it is not possible to calculate E_{disp}).

The first term of eq.(1), E_{es}, corresponds to the classical electrostatic interaction between unperturbed A and B molecules. If we call $\gamma_A^\circ(r_1, R_\beta)$ and $\gamma_B^\circ(r_2, R_\alpha)$ the complete charge distribution (electrons and nuclei) of these two species:

$$\gamma_A^\circ(r_1, R_\alpha) = -\rho_A^\circ(r_1) + \sum_{\alpha \varepsilon A} Z_\alpha \delta(r_1 - R_\alpha) \tag{2}$$

$$\gamma_B^\circ(r_2, R_\beta) = -\rho_B^\circ(r_2) + \sum_{\beta \varepsilon B} Z_\beta \delta(r_2 - R_\beta) \tag{2'}$$

where the same arbitrary coordinate system is employed to describe the position vectors r_1 and r_2 of electrons and R_α and R_β of nuclei, we can write E_{es} as a coulombic interaction energy:

$$E_{es} = \int dr_1 \int dr_2 \, \gamma_A^\circ(r_1, R_\alpha) \, \gamma_B^\circ(r_2, R_\beta) |r_1 - r_2|^{-1} \tag{3}$$

or in an equivalent form which makes explicit use of the electrostatic potential of one of the two molecules (for example A):

$$E_{es} = \int dr_2 \, V_A(r_2) \, \gamma_B^\circ(r_2, R_\beta) \tag{4}$$

with

$$V_A(r_2) = \int dr_1 \, \gamma_A^\circ(r_1, R_\alpha) |r_1 - r_2|^{-1} \tag{5}$$

Expressions (3) and (4) are completely equivalent, but it is well known that it is simpler to get approximate expressions of V and γ than of the two-body interaction operator.

One possible approximation of (4) could consist in replacing $\gamma_B^\circ(r_2, R_\beta)$ with a suitable discrete set of point charges:

$$\gamma_B^\circ(r_2, R_\beta) \simeq \sum_k q_{Bk} \, \delta(r_2 - k) \tag{6}$$

Table I

Schematic definition of the decomposition of $\Delta E_{AB} = E_{AB} - (E_A^\circ + E_B^\circ)$ a) at a given R_{AB} distance

Electrostatic term	$\Psi_1 = \Psi_A^\circ \cdot \Psi_B^\circ$	$E_{es} = E_1 - (E_A^\circ + E_B^\circ)$	Ψ_1 simple product of the separated and unperturbed A and B wfs
Polarization term	$\Psi_2 = \Psi_A \cdot \Psi_B$	$E_{pol} = E_2 - E_1$	Ψ_2 simple product of the separated A and B wfs optimized in the mutual electrostatic field
Exchange term	$\Psi_3 = A\{\Psi_A^\circ \cdot \Psi_B^\circ\}$	$E_{exch} = E_3 - E_1$	Ψ_3 fully antisymmetrized product of the unperturbed A and B wfs
Charge transfer and mixing terms	$\Psi_4 = \Psi_{AB}$	$E_{ct} = E_4 + E_1 - E_2 - E_3$	Ψ_{AB} SCF wf of the AB supermolecule

a) $E_i = \langle \Psi_i | H_{AB} | \Psi_i \rangle$

that, introduced in eq.(4), leads to the simpler expression:

$$E_{es} \simeq \sum_k V_A(k) q_{Bk} \tag{7}$$

The calculation of the coulombic interaction energy (3) is thus replaced by a finite summation of values of a one electron quantity, V_A, calculated at some selected points k and modulated by the values q_{Bk}.

When molecule B is an atomic ion with charge Z_B, expression (7) can be reduced to a single term:

$$E_{es} \simeq V_A(k)\ Z_B \tag{8}$$

there is general agreement to refer to the electrostatic potential under the form of an interaction energy of A with a charge $Z_B = 1$.

The role played by V in the theory of the chemical reactivity consists of the application of equations like (7) or (8) to approximate E_{es} (and, where possible, ΔE_{AB} too). As was said at the beginning, we shall consider applications where V_A is directly obtained through eq.(5) and we shall supplement this discussion with some consideration of a practical method of getting an approximate expression of V_A.

3. LIMITS OF THE ELECTROSTATIC APPROXIMATION

The short resumé given above shows at the same time the justification for the introduction of V in the calculations of chemical reactivity and what are the limits of approximate methods relying only on V.

A separate calculation of E_{es} through eq.(3) (or 4) is noticeably simpler than a complete calculation of ΔE_{AB} and even simpler is the calculation of E_{es} by eq.(7) even if the number of addends in the summation is quite generous.

It is well known, however, that the reactive interaction act cannot be reduced to a classical interaction between rigid charge distributions because other factors, either of classical origin, like the reciprocal polarization of the two partners, or of intrinsic quantal origin, like the exchange effects, play a very important role, which in some cases can be the dominant one.

The relative weights of the contributions to ΔE_{AB} change noticeably when the mutual distance R_{AB} between the two partners changes. Certainly in cases where one molecule at least has a net charge or a permanent dipole moment, the dominant contribution to ΔE_{AB} at large R_{AB} values will be given by E_{es}.

One could thus take the prudent position of considering electrostatic approximations to ΔE_{AB} reliable and useful only for interacting systems of this type and only for large R_{AB}'s.

One can accept without objections the limitation based on the intrinsic properties of A and B, because in most chemical reactions one reactant àt least has a noticeable dipole moment or a net charge, but one cannot deduce reliable information on the really interesting features of a reaction relying solely on large distance calculations.

The essential information on the mechanism of a reaction concerns the identification and the characterization of the transition states and of the intermediates: in some favourable cases this information may be substituted by analogous data concerning larger A-B distances provided that at that stage of the reaction the actual factors characterizing the reaction are already effective.

It might happen that at relatively large distances the interaction energy is dominated by interaction phenomena which do not represent the essential part of the reactive interaction under examination. An actual example can be derived from the study of the mechanism of the basic hydrolysis of the amidic bond we made a few years ago |6|.

The physical model we adopted was composed of a formamide molecule and an OH^- ion: at large distances E_{AB} is well represented by the E_{es} term alone, and the electrostatic energy (which can be safely approximated by $-V_A$, because OH^- is a small ion) indicates the preminence of three approach paths on the molecular plane leading to a direct attack

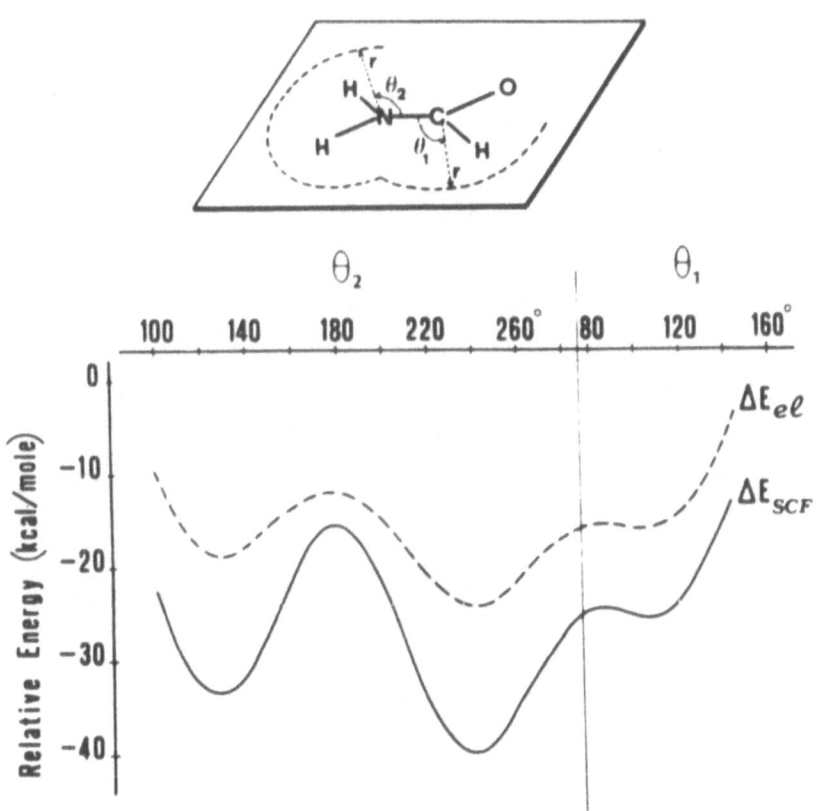

Fig. 1. Trend of E_{AB} (full line) and of E_{es} (broken line) for angular variations of the OH^- position in the formamide plane at r = 2.5 Å.

on the three H atoms, followed by a fourth path, less deep, which corresponds to an attack upon the C atom. The electrostatic description of these paths is shown, in the form of V maps, in figures 19 and 21. The electrostatic approximation is fairly reliable even at intermediate distances, as may be verified by examining figure 1 where a comparison is given between SCF and electrostatic results obtained by placing OH^- on the molecular plane at different positions 2.5 Å apart with respect to the nearer formamide atom. The three planar paths do not correspond to the amidic bond cleavage reaction:

$$OH^- + H_2NCHO \rightarrow NH_3 + HCNOO^-$$

but to other secondary reactions (for example the extraction of a water molecule), and the only path leading to the correct reaction is the fourth one. As a consequence, the information concerning planar paths is not of direct interest for studies on the mechanism of the C-N bond cleavage reaction.

Let us suppose now that we are able to avoid such traps and that the initial portion of the true reaction path (or paths) has been correctly identified. A different objection throws doubt upon the possibility of obtaining sensible information on the reaction characteristics by means of calculations concerning too early a phase of the reaction.

The semiquantitative extrapolations employed in approximate calculations on chemical reactions rely upon more or less clearly stated assumptions on the behaviour of the reaction path with respect to substitutions or conformation changes in the reactant. The empirical assumption more widely employed is the so called "non crossing rule". Such rule states that for similar reactants the ratio of the energies necessary to reach a generic point common to the respective reaction paths is (approximately) proportional to the ratio of the activation energies. It is clear, as one can verify by looking at figure 2, that when such

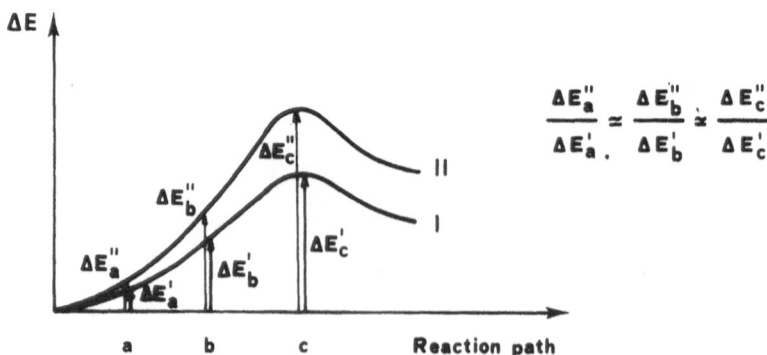

Fig. 2. A schematic representation of two ideal reaction paths obeying the non crossing rule.

ratio is measured at point a very far from the transition state its predictive value is quite modest.

Fortunately experience has shown that in some different classes of reactions the electrostatic approximation is applicable also to relatively small A-B separation. The reasons which permit us to consider E_{es} as a sensible approximation to ΔE_{AB} values are different in different classes of reactions and as a consequence it is necessary to abandon general considerations and to pass to a separate examination of some types of reaction.

4. THE FORMATION OF NON COVALENT MOLECULAR COMPLEXES

The reaction most extensively studied from this point of view is that leading to the formation of hydrogen bonded complexes. This reaction does not present, in vacuo, a barrier, and all the dissections of E_{AB} performed by different authors |7, 8, and references quoted therein| show that E_{es} gives a reasonable picture of the reaction path energy profile from large separations to the dimer equilibrium distance. There is, in fact, a sufficiently good numerical compensation between positive (E_{exc}) and negative ($E_{pol} + E_{ct}$) contributions to ΔE_{AB} which permits the use E_{es} as a fairly good guess of ΔE_{AB}. At R_{AB} values smaller than the equilibrium distance the numerical value of E_{exch} is larger than that of

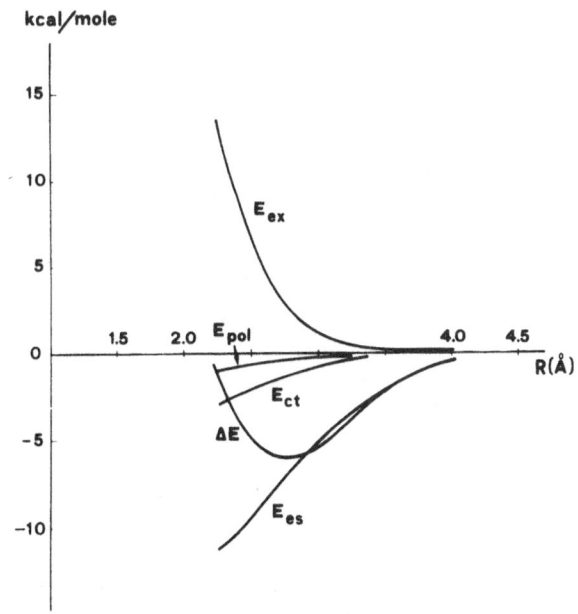

Fig. 3. A typical example of decomposition of ΔE_{AB} for a hydrogen bonded complex (the actual curves refer to the approach of H_2O as a H^+ donor towards the acetamide molecule).

$\left|E_{pol} + E_{ct}\right|$ and E_{es} cannot be used to find the dimer equilibrium separation (we recall that such analyses mainly refer to SCF calculations which neglect the small E_{disp} term).

A typical example is given in figure 3: of course the actual profiles of such curves depend upon the characteristics of A and B and present minor changes due to the basis set employed in the calculations. The numerical results shown in figure 4 indicate the quality of the correlation between ΔE_{AB} and E_{es}, both calculated at the dimer equilibrium distance. Such data are drawn from papers published by Umeyama and Morokuma |7| and by Kollman |9| (4-31G calculations) but they could be supplemented by other numerous analogous results (published as well as unpublished) without altering the quality of the correlation.

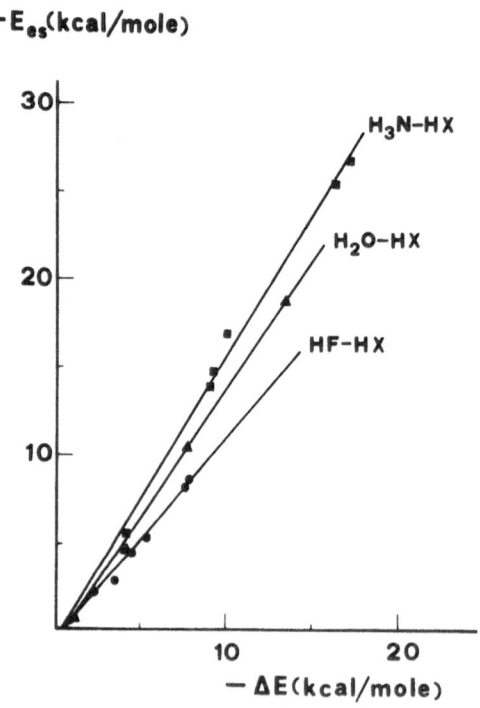

Fig. 4. Correlation between ΔE_{AB} and E_{es} for some hydrogen bonded complexes.

The electrostatic approximation also gives a good representation of the directional properties of the hydrogen bond and of the conformational energy of the dimer. We give in figure 5 a set of three different rotational curves for a $H_2NCHO \cdot H_2O$ complex calculated some years ago in collaboration with M.me Pullman |10| (STO-3G basis set). The full lines refer to SCF calculations on the supermolecules and the interrupted ones to the corresponding electrostatic calculations.

In dimeric (and oligomeric) complexes without hydrogen bond the

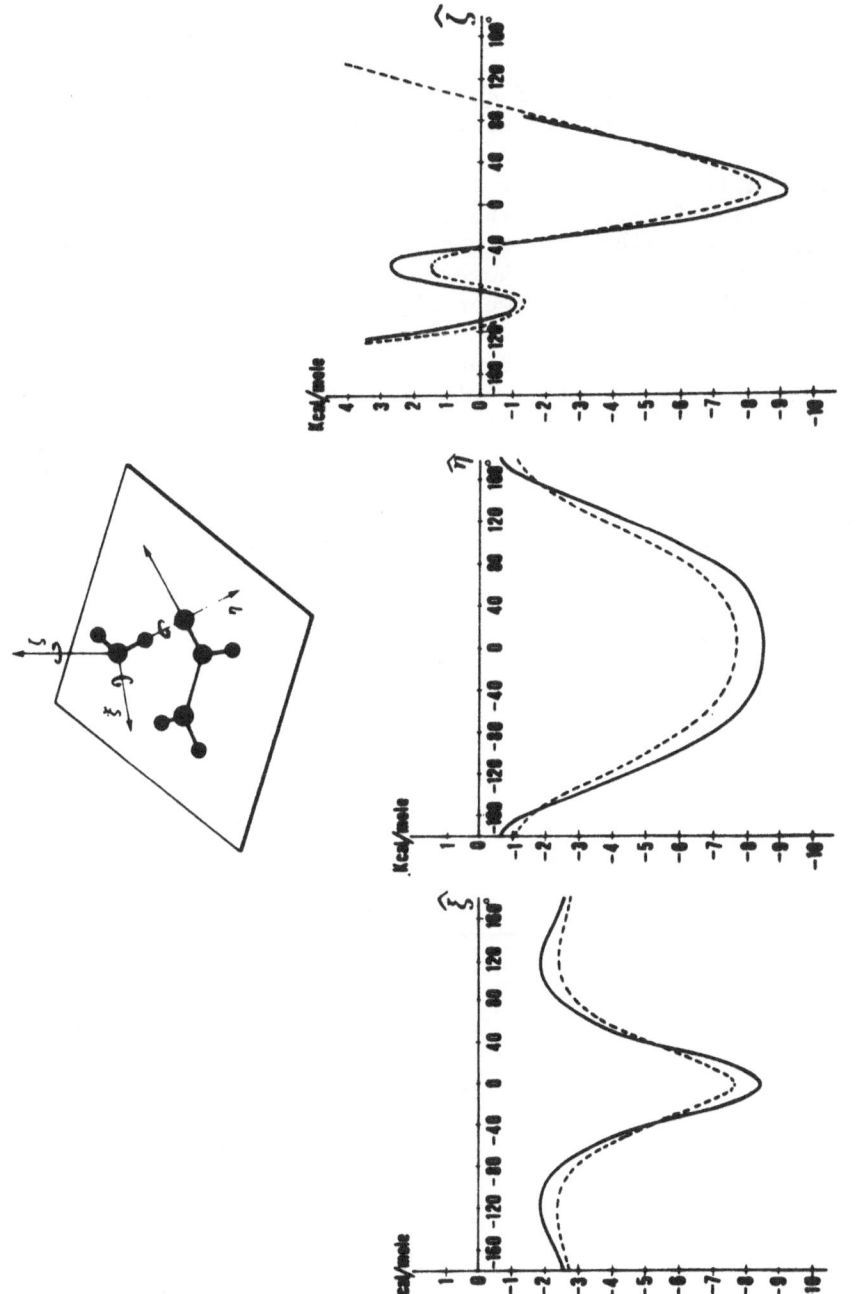

Fig. 5. Conformational curves for the rotation of H_2O with respect to 3 perpendicular axes in a $H_2COH \cdot H_2O$ complex. Full line: ΔE_{AB}, broken line E_{es}.

electrostatic portion of ΔE_{AB} gives again a fairly good representation
of the stabilization energy of the complex as well as of its geometry
(with the exception of the equilibrium R_{AB} value). Recently Kollman |9|
analyzed the formation energy of quite a large number of molecular complexes, ranging from charge transfer complexes to ionic, radical and van
der Waals complexes. Morokuma, in a parallel analysis of the stabiliza-
tion energy of different types of charge transfer complexes |11|, got a
more detailed classification taking into account that for some specific
cases ΔE_{AB} at the equilibrium distance is determined by the combination
of E_{es} and E_{ct}. One may verify, however, that even in such cases there
is a fairly good correlation between ΔE_{AB} and E_{es}.

In all the examples given above the electrostatic term has been calculated by the correct expression of E_{es} (eqs. 3 or 4) without introducing supplementary simplifications. It has been shown that is quite
easy to obtain good point charge representations for γ_B° (especially when
B is a small molecule |12|), but it is evident that in all the molecular
complexes we have quoted the not negligeable dimension of B forbids a direct use of V_A maps to derive quantitative information on the energetics
of the complex.

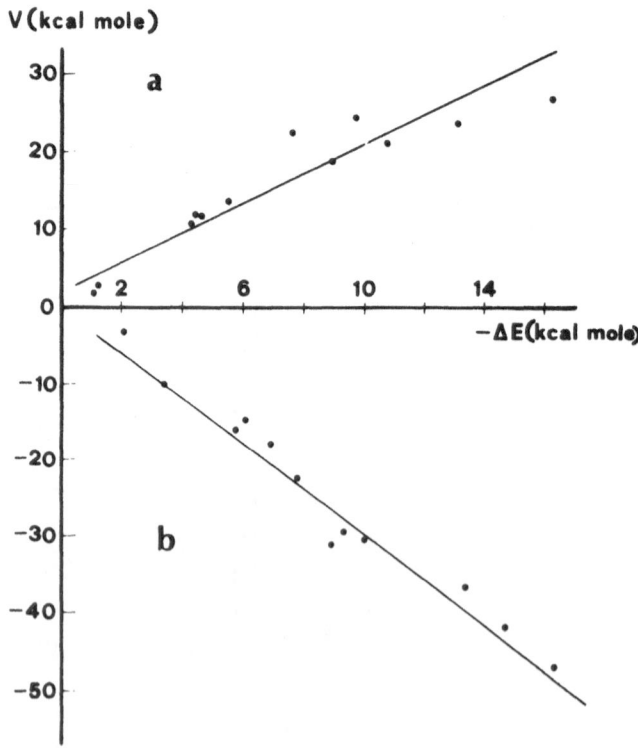

Fig. 6.   Correlation between ΔE_{AB} and V_A (calculated at a specific point) for some H bonded complexes: in curve a) A acts
as a proton donor and in curve b) as a proton acceptor.

A direct use of V_A, determined at a specific point of the space surrounding A, has been indeed proposed by Kollman et al. |13| to obtain an estimate of the stabilization energy of hydrogen bonded complexes. Such authors remarked that in the set of the XH·NH$_3$ and HF·XH complexes the numerical quantity obtainable from the wavefunction of the isolated XH molecule which exhibits the best correlation with the stabilization energy of the dimer is $V_{XH}(k)$, calculated at a prefixed point k. The degree of correlation given by this reactivity index is shown in figure 6. In the case of the XH·NH$_3$ complexes (curve a) the electrostatic potential is measured in a portion of space where V_{XH} is positive (XH acts a proton donor), while in the case of the HF·XH complexes (curve b) the electrostatic potential is sampled in a region where V_{XH} is negative (XH acts as a proton acceptor). The regression coefficient is respectively 0.97 and 0.88. This correlation has been tested for other hydrogen bonded complexes (unpublished calculations) as well as for other non covalent associations |9|.

5. A DIGRESSION ON THE MOLECULAR REACTIVITY INDEXES

This correlation between the formation energy of hydrogen bonded complexes and the value of V at a specific point k permits us to use V(k) as a reactivity index. A short digression on the classification and on the logical status of such indexes may be of some interest.

A reactivity index can be defined as a numerical quantity (which may or may not be a physical observable) obtained from one of the reactants (for example A) which correlates with some characteristic property (barrier height, rate constant, etc.) of the reaction under investigation: A + B → products.

The index can give only comparative indications, i.e. in a set of related reactions, where B is kept constant and A is systematically varied by making appropriate group substitutions (A,A',A'' etc), the changes in that property of the reaction is mirrored by corresponding changes in the index.

The indexes are generally classified as static when they are obtained from the unperturbed molecule and as dynamic when they derive from the wavefunction of the reactant subjected to a schematic perturbation kept constant along the set A,A',A'' etc.

I would like to remark that such distinction between static and dynamic indexes does not imply, "ipso facto", a difference in the quality of the indexes, where the dynamic ones correspond to a more refined approximation level.

It is clear that the final justification for the use of an index and a comparison of the efficacy of different competing indexes derives only from pragmatic considerations, i.e. a molecular quantity is a good index if the correlation it has with the corresponding characteristic of the reaction is good, but it is also evident that its euristic value greatly increases if the index is directly related to that particular feature of the reaction mechanism which determines the characteristic of the reaction under examination.

This question is particularly simple when these characteristics de-

rive directly from the interaction energy (e.g. activation energies or
reaction energies for reactions without barriers). We may introduce an
empirical taxonomy of the reaction types according to what is the term
(or terms) of the decomposition of ΔE_{AB} given by eq.(1) which determines
the change of ΔE_{AB} due to substituent changes in A. The examination of
a few typical cases may better clarify what is the logical status of different reactivity indexes:

1) For reactions where E_{es} is the determinant term, the best indexes
should be of a static type: according to Kollman et al |13| V(k) is a
fairly good index for thys type of reactions.

2) For reactions where E_{pol} is the determinant term, the best indexes should be on the contrary of dynamic type (for example an index
deduced from atomic or bond polarizabilities).

3) For reactions where E_{ct} is the determining term, an assessment
on the pertinent indexes could be obtained by referring to the generalized perturbation treatment of the corresponding interaction energy
term |14-16|. This term can be basically written as

$$E_{ct} \simeq \sum_{m}^{occ} \sum_{\nu}^{uno} \frac{(C_a^m C_b^\nu \Delta\beta_{ab})^2}{E_m - E_\nu} \tag{9}$$

where m are the occupied orbitals of A (in the cases where it is the
electron donor) and ν the unoccupied orbitals of B, C_a^m and C_b^ν the coefficients of the atomic orbitals a and b in the various MO's, $\Delta\beta_{ab}$ is the
interaction integral between a and b, and finally E_m and E_ν are quanti-
ties related to the energies of the various molecular orbitals. Starting
from eq.(9) one can define dynamic indexes by substitution of the quan-
tities related to molecule B with suitable schematic terms (for example
the delocalizability indexes |17| are defined in such a way), but other
indexes, defined in terms of the unperturbed MO's of A (for example the
frontier orbital index |18|) or in terms of the change of the orbital
energies of A, have been proposed and exploited. These two last types of
indexes should be operationally classified as static indexes.

4) The generalized perturbation scheme could be employed also for
examining reactivity indexes for reactions where E_{exch} is the determining
term. For brevity's sake we shall not analyze the indexes resulting from
this approximate expression of E_{exch}, because the results are similar to
those of the preceding case.

5) In other reactions there is not a single term of ΔE_{ab} responsible for the variations of the reactivity along the set of reactants. An
example, taken from a paper of Umeyama and Morokuma |19| is given in fig-
ure 7. The trend of the protonation energy along the set composed by ammonia and its methyl derivatives is given by the combination of E_{es} and
E_{pol}: in this case it is necessary to combine a static (e.g. V) and a
dynamic (e.g. atomic polarizabilities) index to reach the first order
approximation level given by a single static or dynamic index for cases
1) and 2) |20|.

Coming back to the main subject of this exposition, it seems to us

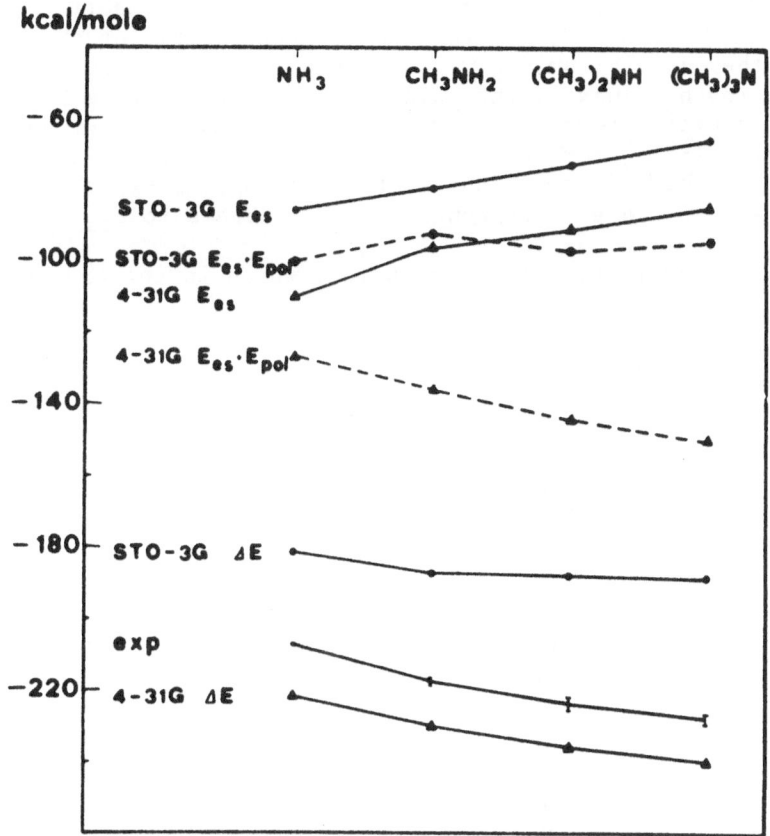

Fig. 7. Trend of E_{es}, $E_{es} + E_{pol}$ and ΔE_{AB} for the protonation of ammonia and its methyl derivatives at the equilibrium proton distance.

that from this digression on the molecular reactivity indexes one could deduce that V can be employed as reactivity index only for classes of reactions satisfying some specific conditions (case 1). If the Kollman analyses |9,13| are accepted, V is the best index thus far proposed for such reactions. In addition V could be employed, in connection with other quantities, to obtain composite indexes having a better predictive value for other classes of reactions (e.g. case 5). These conclusions can be of some help in the examination of the characteristics of other classes of compounds.

6. THE PROTONATION REACTIONS

The example given in figure 7 can be used as an introduction to another class of reactive interactions, that of a proton with a neutral molecule, which has been fairly well investigated.

The interaction energy in the protonation reactions is dominated, at large distances, by the electrostatic term. At shorter distances, near the equilibrium R_{AH} value, the polarization and charge transfer terms reach values of the same order of magnitude as E_{es}, and there is not a positive term which numerically compensates for such contributions to ΔE_{AH^+} (E_{exch} remains quite small). A typical example showing how the ratio of the different contributions change as R_{AB} decreases is given in figure 8. Curve a corresponds to E_{es}, curve b to $E_{es} + E_{pol}$, curve c to ΔE_{AH^+}.

The electrostatic contribution, which in the protonation reactions

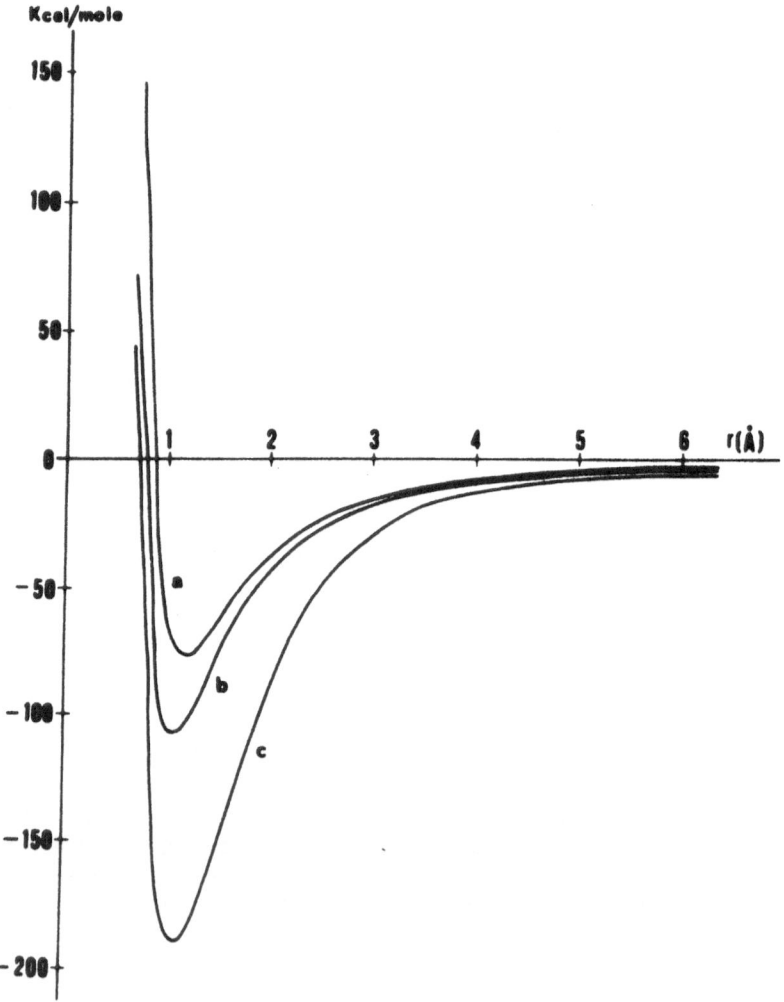

Fig. 8. A typical example of decomposition of ΔE_{AB} for a protonation reaction: curve a: E_{es}, curve b: $E_{es} + E_{pol}$, curve c: ΔE_{AB} (the actual curves refer to the protonation of aziridine.

directly corresponds to V, gives at the equilibrium distance less than a half of the interaction energy, but at the same time gives a fairly correct indication of the final distance of the proton from the molecule. The possibility of using E_{es} to get this information has been checked in many other cases, and even better indications have been obtained on the orientation of the new A-H bond with respect to the molecular skeleton.

In the example of the ammonia derivatives it was found that the changes of the proton affinity are ruled by the changes in $E_{es} + E_{pol}$. This behaviour has been found also in other series of compounds (for example in the first members of the aliphatic ethers family) but is not a general rule. We give as a counter example in figure 9 the correlation

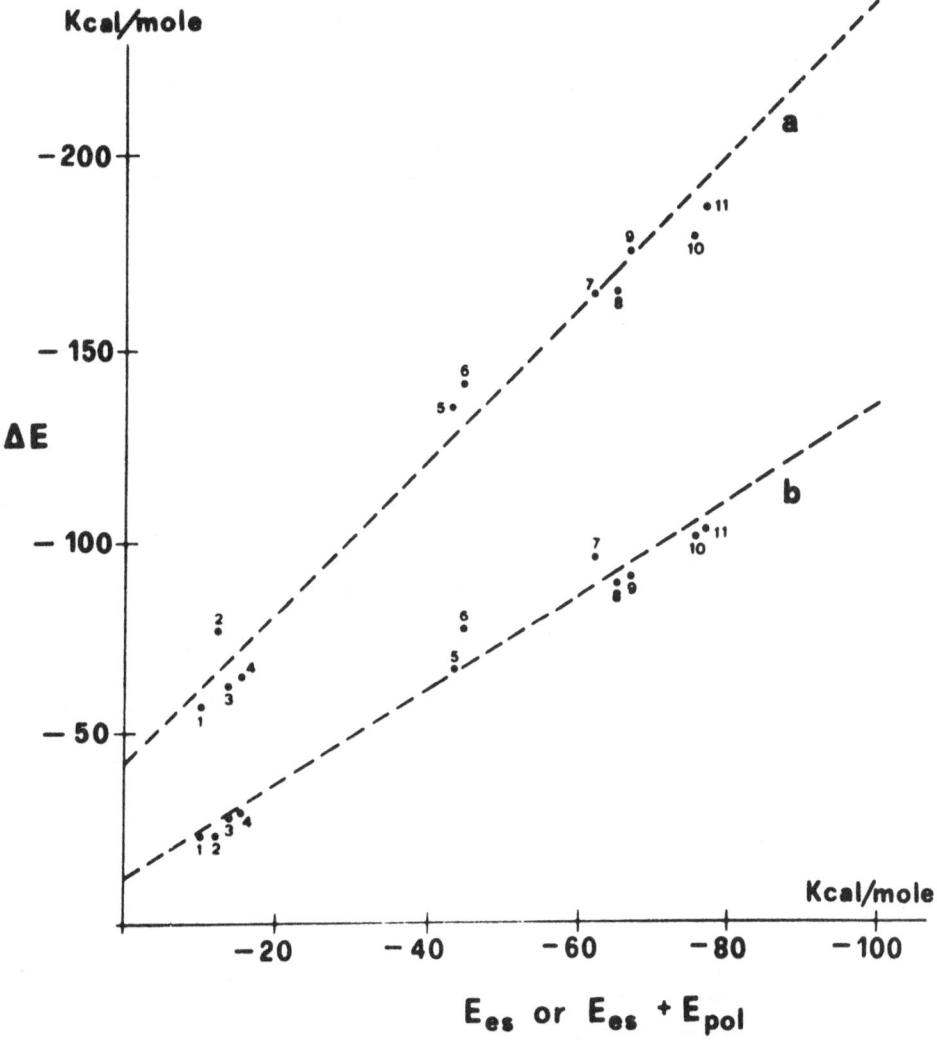

Fig. 9. Correlation between ΔE_{AB} and E_{es} (curve a) and between ΔE_{AB} and $E_{es}+E_{pol}$ (curve b) for the protonation reaction of some three-membered cycles near the equilibrium proton dist.

of E_{es} (curve a) and $E_{es} + E_{pol}$ (curve b) with ΔE_{AB} for a set of protonation reactions in the three membered cycles family |21|: the quality of the correlation which is fairly good in curve a, does not improve noticeably in curve b. A more careful analysis was done some years ago by M.me Pullman |22| for two alternative protonation channels in formamide: it seems that the $P(k)$ function ($P(k)$ is the value of E_{pol} for a proton placed at point k) roughly parallels in that case $V(k)$. This similarity in the very general features of $V(k)$ and $P(k)$ has been verified in some other cases for the regions of the molecular space near different lone pairs of a molecule A, where the approach of a proton is favoured, but is not of general validity for all the outer molecular space |20|. Changes in $V(k)$ and $P(k)$ due to substitutions in a given molecular framework do not necessarily parallel each other. An example is given by the set of methylamines considered in figure 7. The progressive substitution of H atoms with CH_3 groups produces changes in the electrostatic potential in the region near the N lone pair which can be fairly well represented in terms of independent group contributions |23|, whilst at the same time the changes in the polarization term, $P(k)$, in the same region, follow a different trend |19,20|. As we have said above the combination of $V(k)$ and $P(k)$, which can be calculated separately with relatively inexpensive methods, is necessary (and sufficient) to reproduce the trend of the experimental proton affinities in these molecules. It may be added, in parenthesis, that the minimal basis set calculations are not sufficient to represent the correct trend of ΔE_{AH^+}, $V(k)$ and $P(k)$ in this set of reactions |20,24|.

In conclusion, E_{es} may give a reasonable indication of the reaction energy even for the protonation reactions, but the inclusion of the polarization term may give rise to a quantitative refinement of the description, and in some cases also to qualitative changes in the prediction of the trend of the protonation energy in a set of molecules where the number of electrons changes noticeably. More work must be done in order to assess the validity limits of approximations relying solely on $V(k)$.

The interaction of a neutral molecule A with a light metallic cation M^+ (Li^+, Be^{++}, Na^+, Ca^{++} etc.) generally leads to the formation of complexes in which the internal geometry of A is not greatly affected. The partition of ΔE_{AM^+} gives results similar to those already shown for the case of the hydrogen bonded complexes |25|. We report, in Table II, a few examples of the partition of ΔE_{AM^+} near the equilibrium distance. E_{es} is the dominant term and some compensation between terms of differing signs occurs. The examples reported by M.me Pullman in another conference of this Seminar show that such compensation is not always sufficient to permit a forecast of the preferred geometry of the complex (for cases where different sites of A are available for complexation) based on the electrostatic term alone.

The experience thus far gained of these reactions seems to me not sufficient to draw general conclusions: in particular there are insufficient data on the influence of the basis set on the results when large substrate molecules are involved in the interaction. Practically nothing is known about the partition of ΔE_{AM^+} when M^+ is a heavy cation. It is not necessary to emphasize the importance these reactions have in a large variety of fields (industrial catalysis, bio-organic reactions, etc.) and

Table II

Decomposition of the SCF stabilization energy of a few A·M⁺ complexes (in Kcal/mol). [a]

	$H_2O \cdot Li^+$	$NH_3 \cdot Li^+$	$N(CH_3)_3 \cdot Li^+$	$H_2CO \cdot Li^+$	$H_2O \cdot Be^{++}$	$NH_3 \cdot Be^{++}$	$N(CH_3)_3 \cdot Be^{++}$	$H_2NCHO \cdot Be^{++}$
ΔE_{AM^+}	-47.9	-50.7	-45.3	-59.8	-143.7	-164	-191	-197
E_{es}	-51.1	-56.9	-62.7	-52.4	-124.9	-158	-138	-130
E_{pol}	-7.8	-7.0	-23.0	-17.5	-44.7	-49	-110	-119
E_{exch}	12.7	15.1	35.0	11.4	28.4	48	48	35
E_{ct}	-1.7	-1.9	5.4	-1.3	-2.5	-5	8	19
$R_{X-M}(\text{Å})$	1.80	1.96	1.76	1.76	1.59	1.64	1.64	1.48

a) Calculations performed with the 4-31G basis set (5-21G for Li). The values are taken from: P.Kollman, J.Amer.Chem.Soc. 99, 4875 (1977) and P.Kollman, S.Rothenberg, J.Amer.Chem.Soc. 99, 1333 (1977).

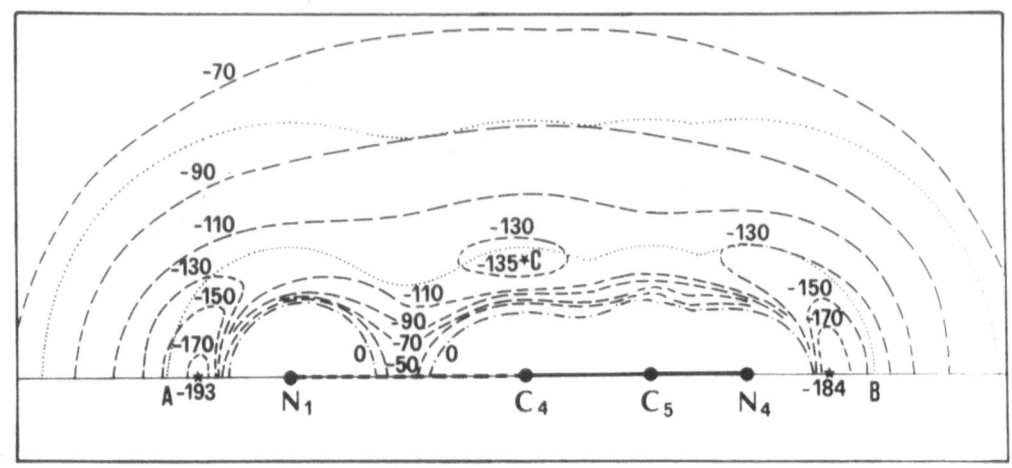

4 CYANO PYRIDINE (4CP) anion

Fig. 10. V map for the p-cyanopyridine anion perpendicular
symmetry plane.

it is quite probable that in the near future the possibility of using
the electrostatic approximation for such reactions will be investigated.

Few data on the partition of ΔE_{AB} for interactions concerning ionic
couples are at present available. The Simonetta's group at Milano |26,27|
showed that the electrostatic approximation gives reasonable results for
the study of the interactions between an organic radical anion and an
alkaline cation. In figure 10 a further example of this kind of application is given. The V map reported in this figure refers to the perpendicular symmetry plane of the p-cyanopiridine radical anion. An examination
of the electrostatic potential, especially on the surfaces lying at a
distance of 1.5 and 3 Å from the atoms of the radical, here represented
by two broken lines, gives a qualitatively correct explanation of the
features of the fine structure of the ESR spectra of this anion in the
presence of Li^+ and K^+ |28|.

7. MORE GENERAL CLASSES OF REACTIONS

We have thus far considered only association reactions, i.e. reactions
of a particular type, where a reaction barrier is not present and where
the internal geometries of the reactants are conserved, within a sufficient approximation, also in the product. The general case of the bimolecular reaction giving products with internal geometry different from
that of the reactants: $A + B \rightarrow C + D$ is surely of a larger interest.

It seems convenient to adopt an empirical classification of the reactions to discuss the possibility of using, in this case too, the electrostatic approximation. The reactions could be divided into three main
categories: the first collecting the reactions where the internal geome-

try of the reactants is conserved in the transition state (or in the product, if the barrier is not present); the second concerning the reactions where there is a specific deformation of one (or both) reactants in the transition state; the third collecting the reactions in which the reaction state is preceded by complex changes in the internal geometry.

For reactions of the first category it is possible to use directly the partition of ΔE_{AB} given in eq.(1). The reactions we have considered so far belong to this category. It may be added that not all the protonation reactions can be put in this class (e.g. the protonation of hydrocarbons).

I shall consider only a couple of example from the second category, sufficient to show the limits and the possibilities of the electrostatic approximation in the study of reactions of this class.

A large part of the aromatic electrophilic substitution reactions pass through an intermediate σ complex, where the deformation of the internal geometry of the aromatic substrate can be schematized by a local deformation concerning only a C atom which changes its hybridization from trigonal to tetrahedral.

It may be of some interest to examine the electrostatic potential of the aromatic substrate in this deformed geometry. (Examinations of maps of V in distorted geometries of aromatic compounds have been already made by Politzer et al.|29|, Politzer and Weinstein |30|, Bertran et al. |31|, using semiempirical wavefunctions). An example is given in figures 11 and 12 where the V maps for the perpendicular plane of benzene in the normal and in the distorted geometries are compared (STO-3G calculations). The map of normal benzene is characterized by two symmetrical regions where V is negative. In the second map the change of hybridization of the C_1 atom brings the C_1H bond below the ring plane and the symmetry of V is destroyed. The shape of V shows that this deformation of the geometry produces a noticeable increase of the E_{es} term for the approach of an electrophilic reactant (whether ionic or dipolar) towards the C_1 atom.

In the specific case of protonation (which can be considered as the first step in a hydrogen exchange reaction) a decomposition of ΔE_{AH^+} in this deformed geometry gives results quite similar to those already discussed for the protonation of a lone pair: the prediction given by E_{es} alone of the final geometry is good, but the electrostatic term corresponds only to a relatively small portion of the interaction energy, and the largest contribution is given by E_{pol}.

In a study of the reaction coordinate for the simplest case of Friedel-Crafts reaction, the hydrogen exchange in benzene induced by HF in presence of BF_3 as catalyst |32|:

$$C_6H_5D + HF + BF_3 \rightarrow C_6H_6 + DF + BF_3$$

we verified that the electrostatic portion of the interaction energy, calculated at this deformed geometry of C_6H_6, gives a fairly good representation of the reaction path near the transition state.

The same electrostatic approach has been employed in the preliminary phases of a not yet completed study of the Gatterman reaction on

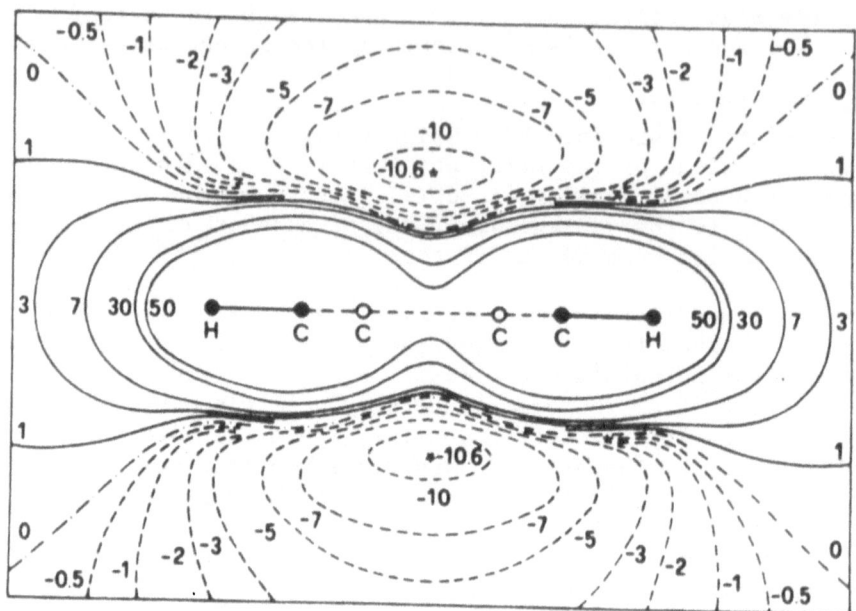

Fig. 11. V map for a perpendicular symmetry plane of benzene.

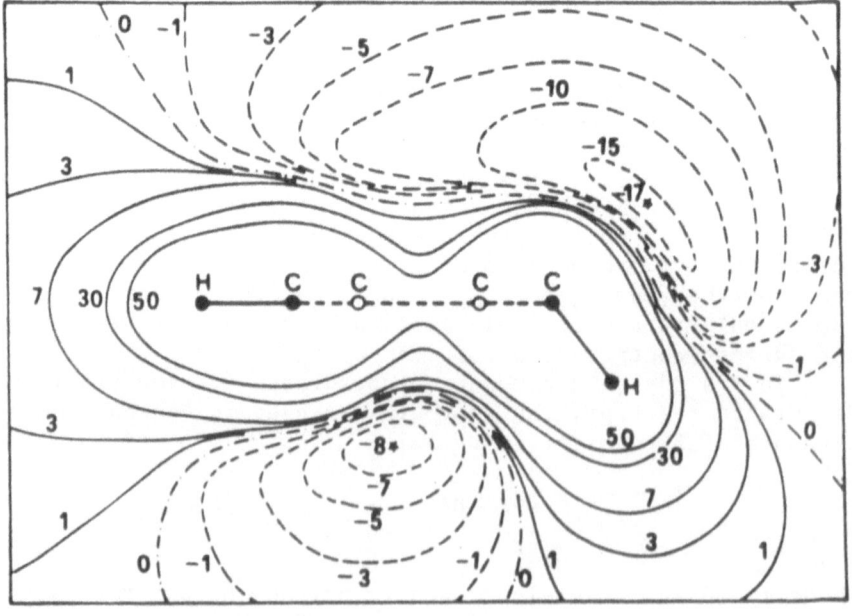

Fig. 12. V map for the same plane of fig. 11 of benzene in a
distorted geometry.

the benzene substrate:

$$C_6H_6 + HCN + HCl \rightarrow C_6H_5CH = NH_2^+Cl \rightarrow \text{products}$$

In this reaction the stoichiometry and the structure of the attacking reagent are still subjects of discussion. The electrostatic approximation seems to be able to discriminate between more and less probable structures |33|, but more precise numerical checks are necessary to get a definite evaluation of the validity of this use of the electrostatic approximation.

It would be interesting to verify if the changes in the electrostatic potential at a deformed geometry due to the introduction of substituent groups in the benzene molecule parallel the corresponding changes in geometry. This investigation has not yet been performed, but an analogous problem is being studied at present in our Laboratory for a different substrate and a different reaction. In the reduction reaction of the carbonyl group of cyclopentanone by metal hydrides the presence of substituents made non equivalent two approach paths (upon and below the ring plane) and the changes in the stereospecificity of the reactions could be in relation to the corresponding changes of V either at the normal geometry (early stages of the reaction) as well as at a deformed geometry having the C atom of carbonyl in a pyramidal conformation (portion of the reaction path near the intermediate complex). In this case too the electrostatic results seem to be in accordance with the experimental evidence |31| but it is too early to draw any conclusions.

A second class of reactions where it seems relatively easy to put in evidence a specific deformation of the substrate is that of the nucleophilic substitution reactions proceding via a S_N2 mechanism:

$$X + \;\; \underset{\diagdown}{\overset{\diagup}{C}} - Y \;\; \rightarrow \;\; X \cdots \overset{|}{\underset{\diagdown\!\diagup}{C}} \cdots Y \;\; \rightarrow \;\; X - \underset{\diagdown}{\overset{\diagup}{C}} + Y$$

One may assume that in the transition state the C group is planar. The classical example of this class of reactions is given by the Walden inversion of methane and substituted methanes, which has been the subject of accurate quantal investigations |35|.

We give in figure 13 the V map for CH_4 and CH_3F calculated at this deformed geometry (4-31G calculations). These calculations have been performed solely for this discussion and are not supplemented by the calculations on the corresponding supermolecules necessary to make a dissection of ΔE_{AB}. A qualitative evaluation of the trend of V_A at some selected points of the two maps of figure 13 and of a few other analogous maps for other substituted methanes seems to be in general agreement with the trend of the interaction energies deduced from preceding calculations performed with different basis sets and with different constraints on the internal geometry. This qualitative evaluation is of course quite imprecise and this example can be considered, at present, nothing more than a suggestion of a possible subject for investigation.

This introduction of a distorted geometry, which is kept constant for a portion of the reaction path, can give rise to completely artificial alterations in the description of the energy profile along the re-

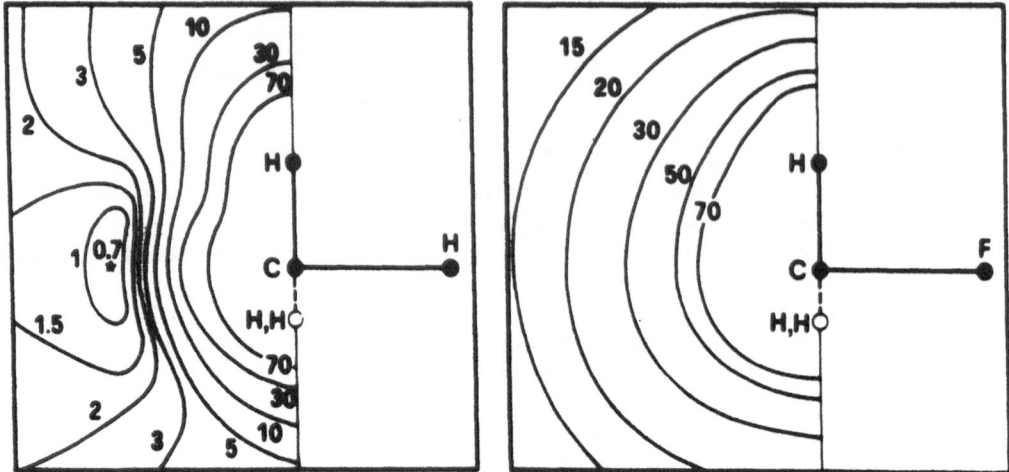

Fig. 13. V map for CH_4 and CH_3F in distorted geometries.

action path. This approximation must be considered no more than an attempt to carry out a separation of effects which actually occur at the
same time. Its euristic value is not completely established, and it is
possible that the schematization we have outlined here will prove too
naive.

The information one may derive from partitions of ΔE_{AB} at arbitra-
rily distorted geometries (and supplemented by the results of analogous
decompositions is performed on the supermolecule at some crucial points
of the reaction path, preceeded by a localization of the MO's on the A
and B portions of AB, such as those presented in this Seminar by the
Leroy group |36|) could however lead to the elaboration of simplified
methods for the calculation of ΔE_{AB} also in the cases where the distortion plays an important role. A discussion of the potential interest of
such methods and of the suggestion that in some cases a description limited only to the classical terms (polarization and electrostatic) is
possible, goes beyond the limits imposed on this paper and must be de-
ferred to another occasion.

8. PHOTOCHEMICAL REACTIONS

The separation of a portion of the reactive process is more acceptable
in photochemical reactions. The photon absorption act, the electronic
rearrangements following the absorption (and in some cases also the corresponding nuclear rearrangements) can be considered separately and the
study of the photochemical reaction can be limited to the second member
of this pair of sequential reactions:

$$A + h\nu \rightarrow A^* \tag{10}$$

$$A^* + B \rightarrow A^*B \rightarrow products \tag{11}$$

Before passing on to discuss the possibility of using the electrostatic approximation for reactions of this kind we have to examine
briefly a preliminary problem, namely the dependence of V_A, ΔE_{AB} and its
decomposition on the basis set and on the method adopted for the calculation of the wavefunctions of A^* and A^*B. This argument was not introduced before, because the experience thus far gained on ground state systems is widely known and easily available elsewhere. Less is at present
known about reactions including an excited state reactant, and a few remarks on this subject may be of some interest.

Some comparisons on the shape of V_A* obtained with different methods of calculation of the wavefunction and with different basis sets for
the ground, the $^3A_1(\pi \to \pi^*)$ and the $^1A_1(\pi \to \pi^*)$ states of formaldehyde can
be found in a recent paper of Daudel et al. |37|. Different method (SCF
and extensive CI for the ground state, rigid excitation, CSECI, EHP and
extensive CI calculations for the excited ones) have been employed in
combination with different bases (STO-3G, STO-3G+sp, 4-31G, 4-31G+sp).

From the results of this investigation, and from analogous unpub-
lished results concerning other states of H_2CO and other molecules having
similar chromophores, one can draw some provisional conclusions.

The promotion of one electron to a virtual SCF orbital is never a
good approximation, the complete single excitation configuration interaction (CSECI) gives a reasonable representation of V_A* (if compared to
that obtained by a large CI) which is however nearly equivalent to that
obtainable at less cost with the electron-hole potential approximation
(EHP) proposed by Morokuma |38|.

The minimal basis set (STO-3G) seems to give qualitatively correct
results, and the implementation with a set of diffuse s and p orbitals
(STO-3G+sp basis) gives results in fairly good accordance with a larger
basis (4-3G+sp). These conclusions are not valid for the $^1(\pi \to \pi^*)$ state
of formaldehyde nor, probably, for analogous states of similar chromophores. To permit a direct evaluation of some of these conclusions we
show in figure 14 some maps of V, calculated with the 4-31G+sp basis for
the perpendicular plane of H_2CO, where the differences among the various
methods are more evident.

The dependence of ΔE_{AB} and of its decomposition on the basis set
was examined by Iwata and Morokuma for the complexes $H_2CO^* \cdot H_2O$ and for
similar complexes involving other organic molecules |39|. In all cases
E_{es} fairly parallels $\Delta E_{A}*_{H_2O}$ at the final geometry: the ratios of the
different terms of the decomposition of $E_{A}*_{H_2O}$ show a dependence on the
basis set not too different from that found in the corresponding ground
state associates. Even if the information available on excited state reactions is much less than that on ground state systems it seems that the
conclusions of the preceding discussion can be adopted as a starting
point for investigations on these reactions.

The electronic excitation produces remarkable changes in the shape
of V. Because there are not many examples of maps of V in the literature,
it will be instructive to take a look at a few examples.

In the figures 15-16 the in-plane maps for the ground and the $(n \to \pi^*)$
states of nitrosoamine (H_2NNO) are shown. The electronic excitation produces the complete deletion of the negative zone near the O atom.

The promoted electron is deplaced on the N-N group: this deplacement

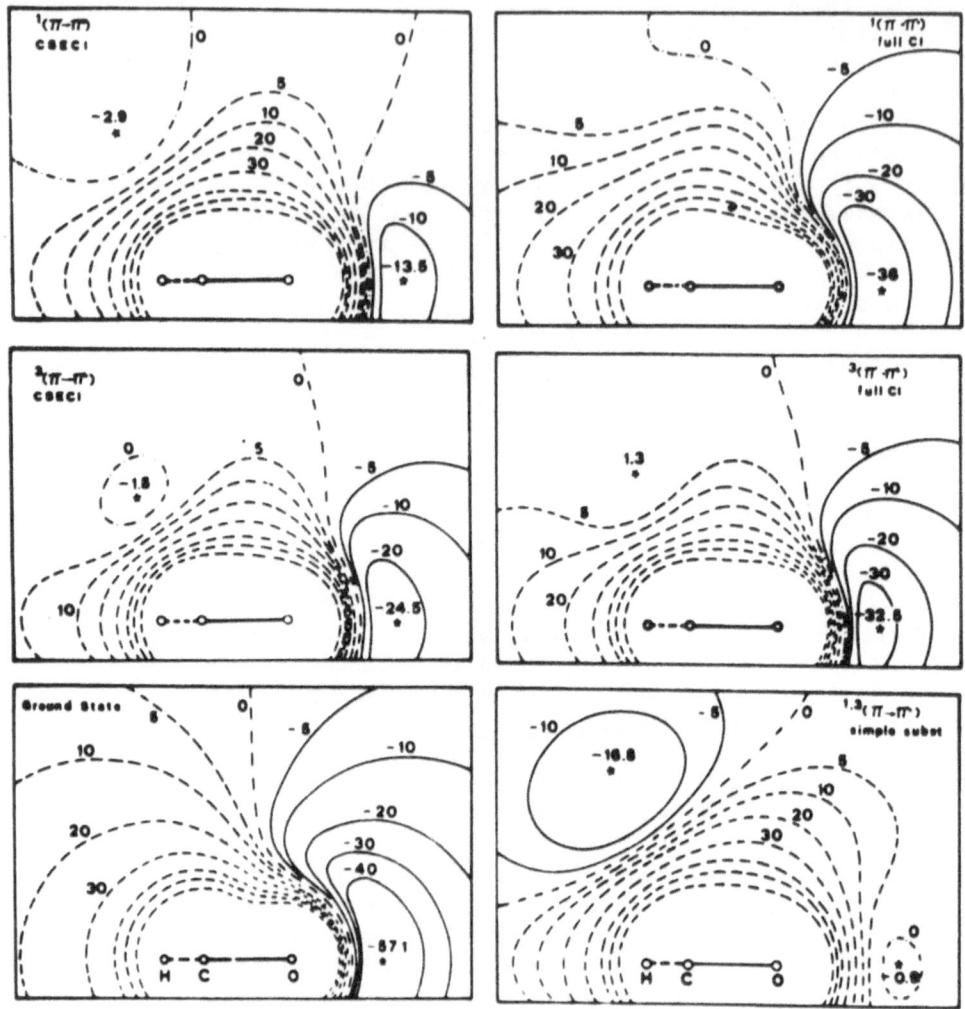

Fig. 14. V maps for the perpendicular symmetry plane of H_2CO and referring to the ground state and to different approximations of the $^1(\pi \to \pi^*)$ and $^3(\pi \to \pi^*)$ states.

can be appreciated by examining the maps of the perpendicular planes passing for the N-N and N-O bonds (figures 17-18). In the family of compounds having H_2NNO as a progenitor it is quite easy to get a photodissociation in slightly acidic conditions (with the formation of NO and HNR_2^+) while the dissociation occurs with more difficulty in neutral or strongly acidic media |40|. We cannot examine here the details of this mechanism, and it is sufficient for our purposes to mention that the shape assumed by V in the two states gives useful information on some important aspects of the different behaviour of this substrate in different media under the action of light.

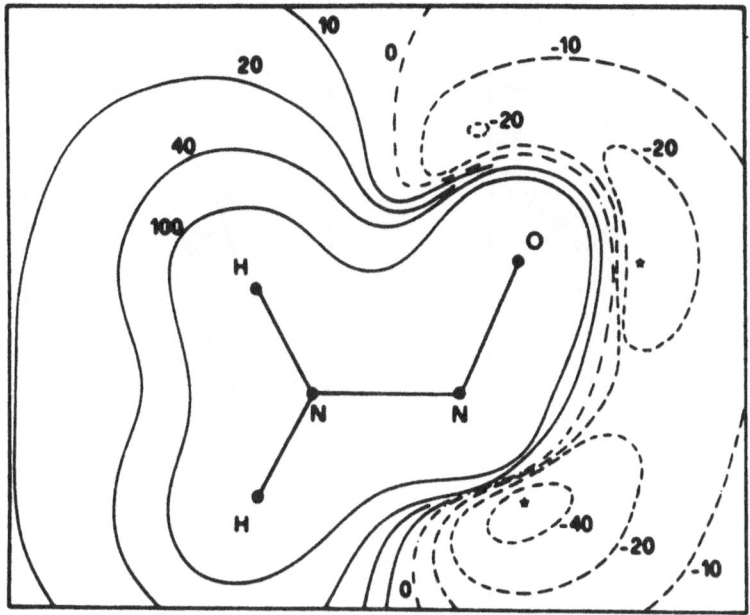

Fig. 15. V map for the molecular plane of nitrosoamine (ground state).

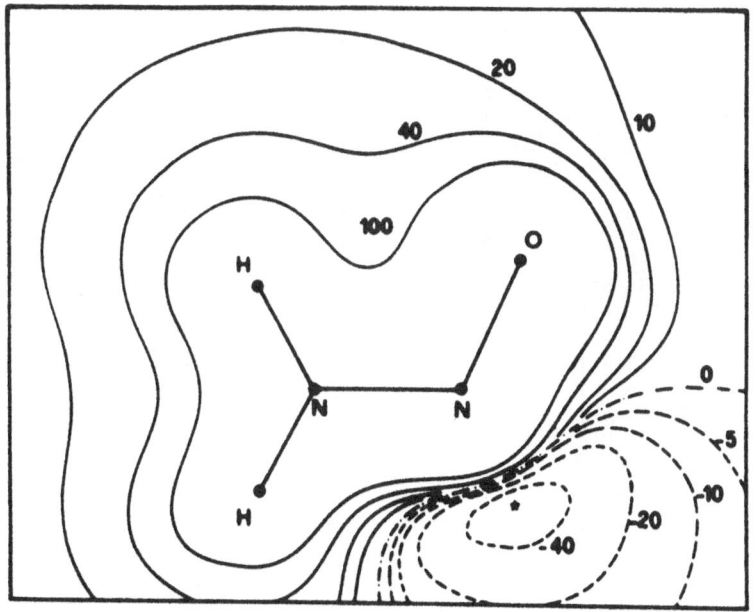

Fig. 16. V map for the molecular plane of nitrosoamine ($^1(n \rightarrow \pi^*)$ state).

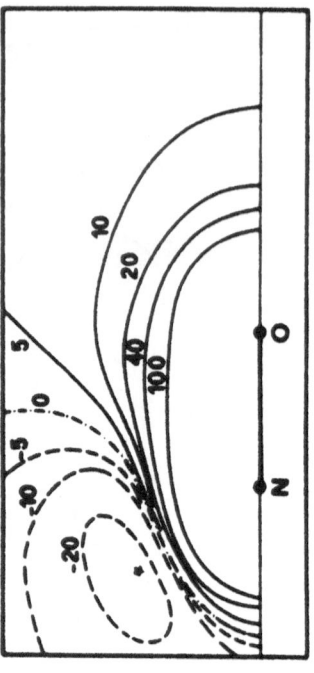

Fig. 18. As fig. 17 for another perpendicular plane containing the N-O bond.

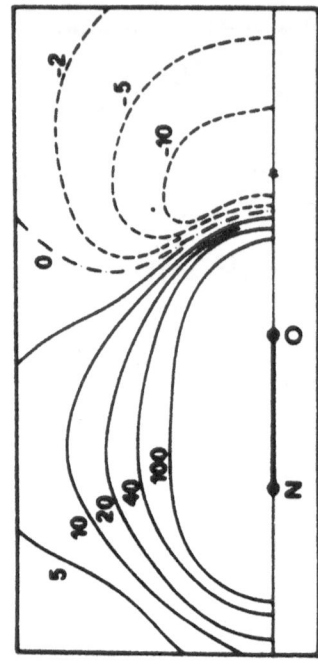

Fig. 17. V map for the perpendicular plane of H_2NNO containing the N-N bond: a) ground state, b) $^1(n{\to}\pi^*)$ state.

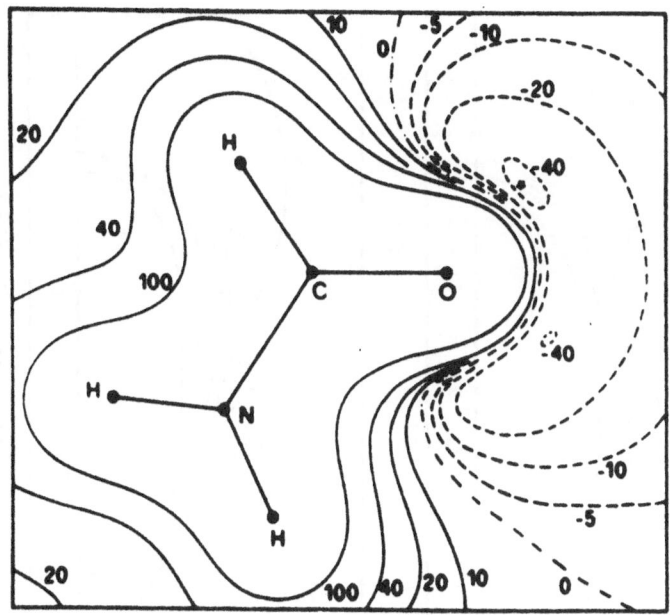

Fig. 19. V map for the molecular plane of formamide (ground state).

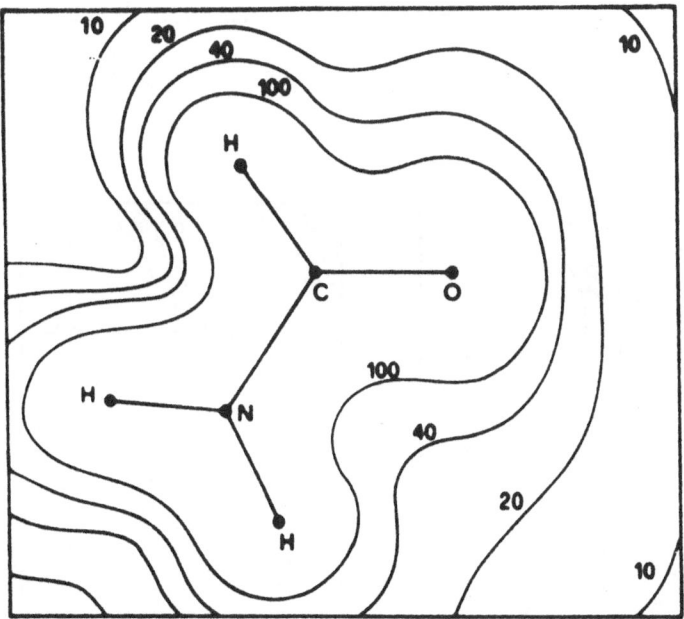

Fig. 20. V map for the molecular plane of formamide ($^1(n \to \pi^*)$ state).

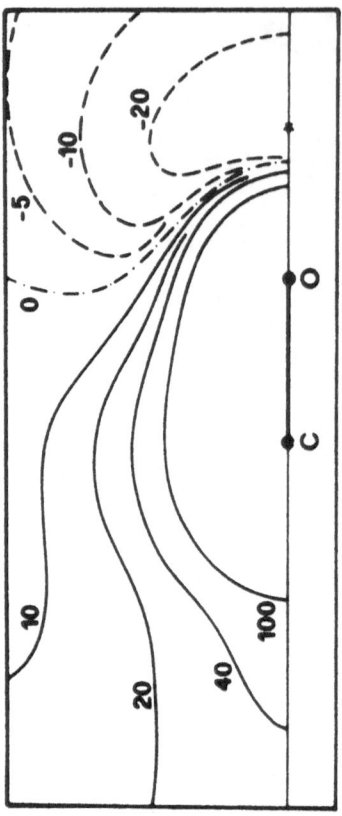

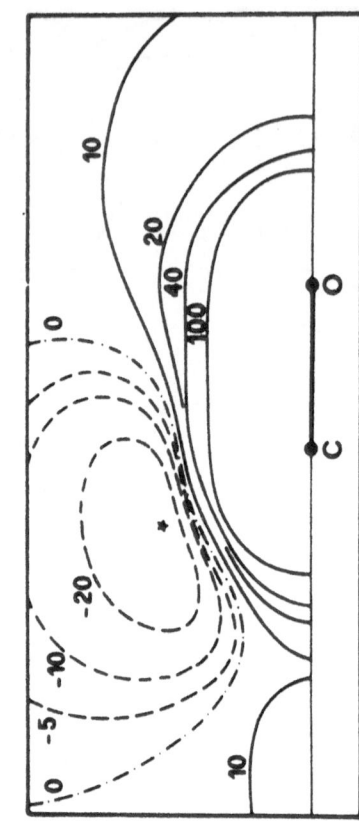

Fig. 22. As fig. 21 for the perpendicular plane containing the C-O bond.

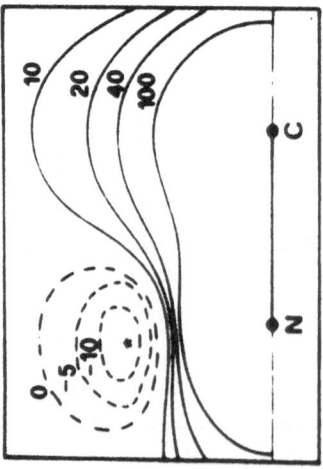

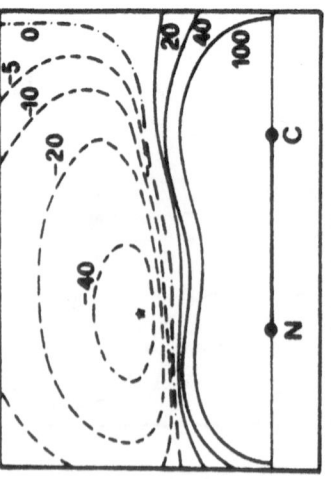

Fig. 21. V maps for the perpendicular plane of formamide containing the N-C bond: a) ground state, b) $^1(n{\to}\pi^*)$ state.

It was noted same time ago 23 that one may get a rationale of the shape of V in terms of nearly transferable group contributions. No attention has been thus far paid to V for excited states from this point of view, but the next example will show that an analysis of this kind is possible. The formamide molecule differs from the preceding one by having a CH group in the place of a N atom. The difference between the in-plane maps of V for the ground state of H_2NNO (figure 15) and of H_2NCHO (figure 19) are easily understandable in terms of additive group contributions. If one compares the maps for the $^1(n_o \to \pi^*)$ state of formamide (which is not the lowest excited state) given in figures 20,21,22 with the corresponding maps of the nitrosoamine molecule (figures 16-18) one can verify that the excitation mechanism is quite similar in both cases. The hole is formed in the O lone pairs and the particle is well localized on the N-C (or N-N) bond. From this example (and others which can be found, e.g., in ref. |41|) one could infer that a rationale for the shape of V_A^* could be made in terms of localized group contributions supplemented by hole and particle contributions.

It should be noted that the possibility of representing V in terms of separate and nearly transferable contributions does not correspond to saying that the examination of the shape of V does not give anything more than chemical insight suggests. The juxtaposition of group contributions is rather subtle and it is quite easy to find examples where the maps of V do not correspond to a simple application of chemical insight.

As an example drawn from the results of a forthcoming publication |42| we may quote the case of the electrostatic potential, in the ground state, of $H_2C=CH_2$, $H_2C=CHCN$ and substituted compounds, where the presence of a nearly localized π bond does not always correspond to the presence of a negative region. Such cases will be probably even more frequent in excited state molecules. The domain of the possible photochemical reactions is still largely unexplored and even less is known on the possibility of using V as an aid in the study of these reactions. There are however a few indications that in some specific classes, for example the photoaddiction reactions, some interesting information can be drawn from maps of V |43|.

9. APPROXIMATE, AND ECONOMIC, METHODS FOR CALCULATING V

We have thus far examined a set of cases where the electrostatic potential might constitute a more or less important implement for studies on chemical reactivity. The interest of this approach is tempered by the not negligeable computational effort necessary to known $V_A(k)$ at an adequate number p of points. In fact one has to add the time necessary for the calculation of Ψ_A to the time for computing p times V_A (i.e. a one-electron observable). This time is only a small fraction of that necessary for complete calculations on the supermolecule, but for large and very large molecules, such as those of interest in biochemical processes, it remains quite exacting.

We think it may be possible to obtain fairly good approximate results with a drastic reduction in the computation time, which depends only linearly on the number n of electrons. I will summarize here the approach we have followed.

The complete molecular charge distribution $\gamma_A^o(r_1)$ (eq.2) can be expressed in terms of quasi-localized neutral contributions $\gamma_{Ai}(r_1)$, if one replaces the canonical orbitals φ_u with the localized ones λ_i and makes a suitable partition of the nuclear charges among the terms $\lambda_i^o(r_1)$ $\lambda_i^*(r_1)$ of ρ_A^o. This substitution having been made, one has:

$$V_A(r_2) = \sum_{i=1}^{n} V_{Ai}(r_2) = \sum_i \int dr_1 \; \gamma_{Ai}^o(r_1,R_\alpha) \; |r_1 - r_2|^{-1} \qquad (12)$$

with

$$\gamma_{Ai}^o(r_1) = -2\gamma_i^o(r_1) \; \gamma_i^{o*}(r_1) + \sum_{\alpha=1}^{2} \delta(r_1-R_\alpha) \qquad \alpha\varepsilon\lambda_i \qquad (13)$$

The last summation in eq.(13) means that if λ_i is a lone pair on atom μ there will be in λ_{Ai} a charge +2e at position R_μ; similarly if λ_i is an orbital bonding atoms μ and ν there will be two unit positive charges at positions R_μ and R_ν, etc.

Expression (12) is of course equivalent to the original definition of V but permits the introduction of some simplifications. The first simplification consists of replacing each λ_i^o with its main part $\bar\lambda_i$ (i.e. the localized orbital withouts tails and renormalized). The second in ignoring the complications deriving from the calculations of a one-electron observable (i.e.V) in terms of non orthogonal one electron functions. The third consists of the substitution of each $\bar\gamma_{Ai}^o$ with a "model" γ_i obtained by making the arithmetic mean of different $\bar\gamma_{Ai}^o$ of the same kind (for example C-H bonds, O lone pairs, etc.) issuing from a few suitably selected molecules.

These approximations might seem very drastic, but an examination of the results will show that they work fairly well.

We give in figure 23 a comparison of two maps of 2-formylamino acetamide (in its fully extended conformation) obtained a) via a full SCF calculation and without any further manipulation, b) via the transferable models $\bar\gamma_i$. The accuracy of the approximate representation, which is sufficient for all the practical applications we have considered in the preceding pages, is retained in other conformations of the same molecule as well as in different molecules.

The reduction of the computation time is remarkable since the time needed to get the wavefunction is now zero (the SCF charge distribution is now replaced by the juxtaposition of rigid models which are directly transferable) and the time needed to obtain the V map is noticeably reduced because the functional space is partitioned into subspaces having only small intersections.

It may be remarked in passing that the substitution of the "ab initio" SCF wavefunction with a semiempirical one requires a larger computation time (the time for the calculation of Ψ, albeit small is not equal to zero, and the time needed to obtain the map is practically equal to that necessary for the SCF calculation $|44,45|$) and gives results no better than the transferable models. We give in fig.24 the CNDO map of 2-formylamino acetamide in the same conformation as that considered in

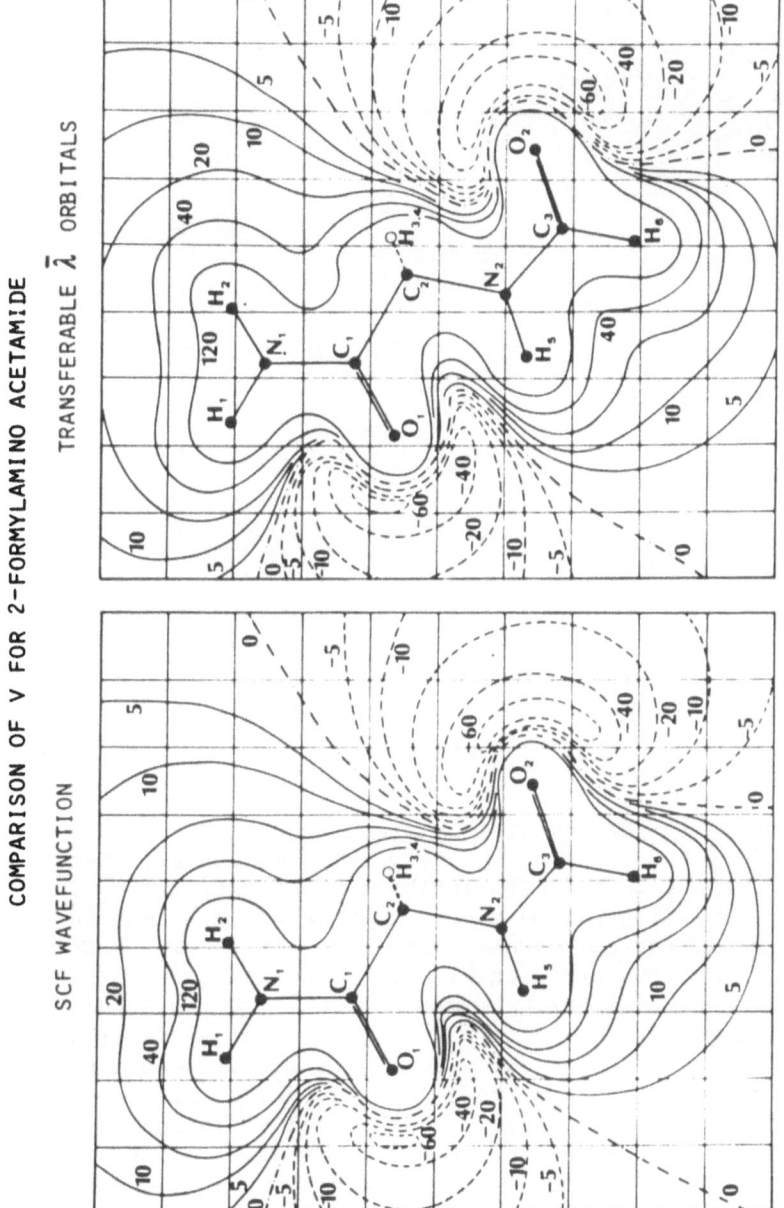

Fig. 23. A comparison of the V maps for the molecular plane of 2-formylamino acetamide calculated by using: a) the SCF wavefunction, b) the transferable group contributions.

CNDO WAVEFUNCTION

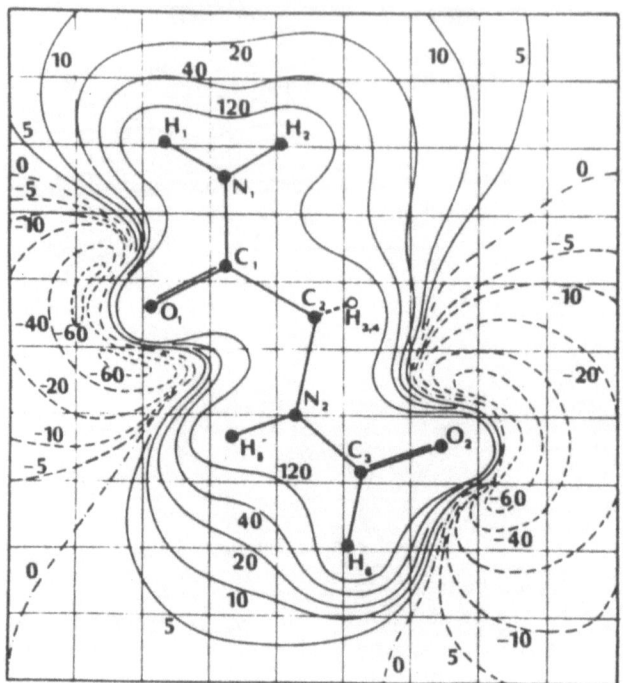

Fig. 24. V map for the molecular plane of 2-formylamino
acetamide calculated by using the CNDO wavefunction.

figure 23.
     The reduction in the computation times we have thus far obtained is
not yet sufficient for the study of very large molecules. There is however the possibility of introducing further approximations.
     The LCAO expression of the models $\bar{\lambda}_i$ without tails can be replaced
by a suitable point charge representation. To obtain good charge representations it is convenient to examine the electrostatic potential $\bar{V}_i$ of
the corresponding neutral charge distribution $\bar{\gamma}_i$, because the potential
maintains a continuous shape also for discrete charge distributions. For
some groups of more common occurrence we have optimized the point charge
distribution by a fitting on quite a large number of positions placed
around the group itself and outside the van der Waals spheres. Some re-
sults can be found in a paper by Bonaccorsi et al. |46|. Point charge
distributions composed of two negative charges for a bond (of sigma or
of banana type) and a negative charge only for a lone pair are sufficient
to represent V with good accuracy outside the van der Waals surface of
the molecule.
     By using this approximation it would be possible to represent the
electrostatic potential of an entire molecule having 2n valence electrons
(the inner shell electrons can be discarded) by a number of negative

charges lying between 2n and n. We have already remarked that the optimization of the point charge distributions was made on points lying outside the van der Waals surface of the group, while the electrostatic potential of points lying just inside such surface plays an important role in some specific aspects of the electrostatic behaviour of the molecule (location of the minima of V, electrostatic stabilization energy of H-bonded complexes, etc.). We have not found useful to obtain an optimization of the point charge distributions valid also for a region inside the van der Waals surface, because the number of charges necessary for the required precision increases rapidly, and we suggested a different solution, namely to keep the LCAO expression of the $\bar{\gamma}_i$'s lying immediately next to the region where V is sampled and to use the point descriptions for the other groups.

For example we have found that a reliable representation of V near a carbonyl group may be obtained by using only the LCAO expression for the models of the O lone pairs and for the two banana C-O bonds. The electrostatic potential obtained with this formula for the region near a carbonyl of 2-formylamino acetamide is given in fig. 25a. It can be noticed that this map is practically equal to that given in fig. 23b.

It is convenient to point out that the quality of the representation of V is quite sensitive to the description of the groups lying far apart. We show in figure 24b the same portion of the map of 2-formylamino acetamide obtained by again using the LCAO expression of the carbonyl group and replacing the point charges of the other ones by what we may call "transferable atomic charges" i.e. negative charges placed on the nuclear and obtained from the LCAO expressions of the $\bar{\gamma}_i$ by the same rule as the Mulliken populations (of course one cannot employ here the true Mulliken populations, because there is not a wavefunction of the whole molecule). The map of figure 25b shows noticeable deviations from that of figure 23b.

Recently Hayes and Kollman |47| obtained the first example of a 3D representation of the electrostatic potential of an actual protein, the carboxypeptidase A and from this description of V a first approximation of the catalytic mechanism of this enzyme. These authors used in this study a point charge representation of subunit transferable models which resembles our transferable atomic charges. The size of carboxypeptidase A is by far larger than that of 2-formylamino acetamide, and it is quite probable that this approximation may be profitably employed for groups lying at larger distances from the portion of space where V is sampled.

Returning now to the representation of V in terms of a few LCAO transferable models and of the point charge distributions we have already introduced, it may be added that this approximation was found sufficient to reproduce the trend of the energy of formations of complexes with water in a fairly large number of cases (a few are given in ref. |46|) and to give adequate representations of the corresponding conformational curves.

An example of these curves is given in figure 26 and refers to a complex between water and the C=O group of 2-formylamino acetamide already considered. Curve a refers to SCF calculations, curve b to calculation with LCAO models of C=O and point charge models for the other groups, curve c to calculations with LCAO models for C=O and transferable

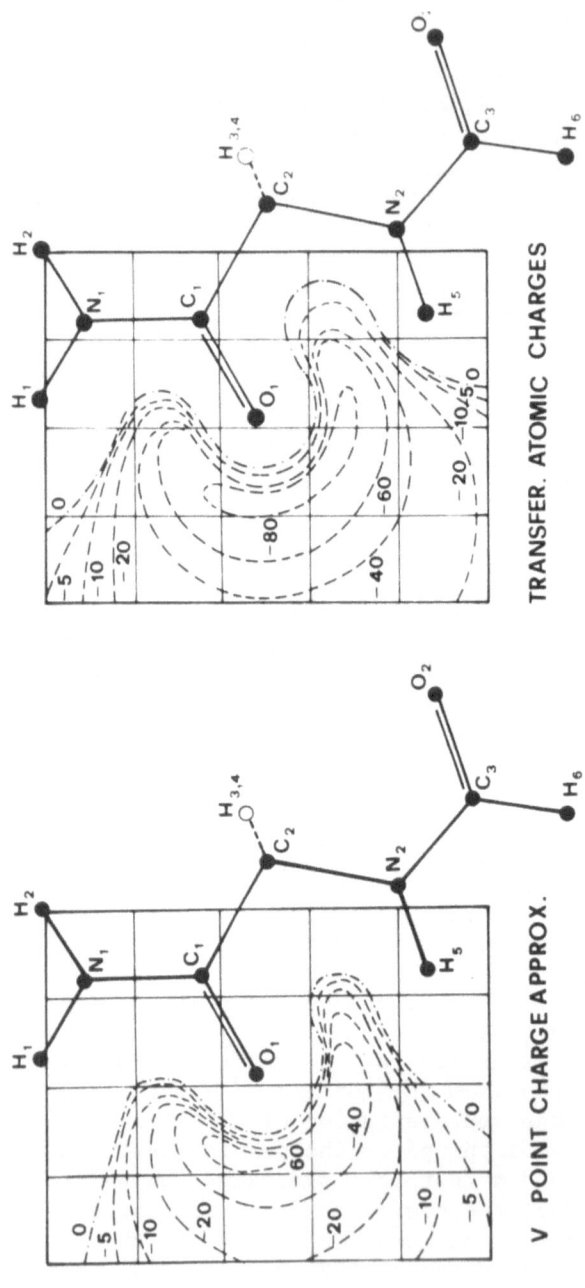

Fig. 25. A portion of the V map for the molecular plane of 2-formylamino acetamide calculated with two different approximations.

CONFORMATIONAL ENERGY FOR A COMPLEX OF 2-FORMYLAMINO ACETAMIDE AND WATER

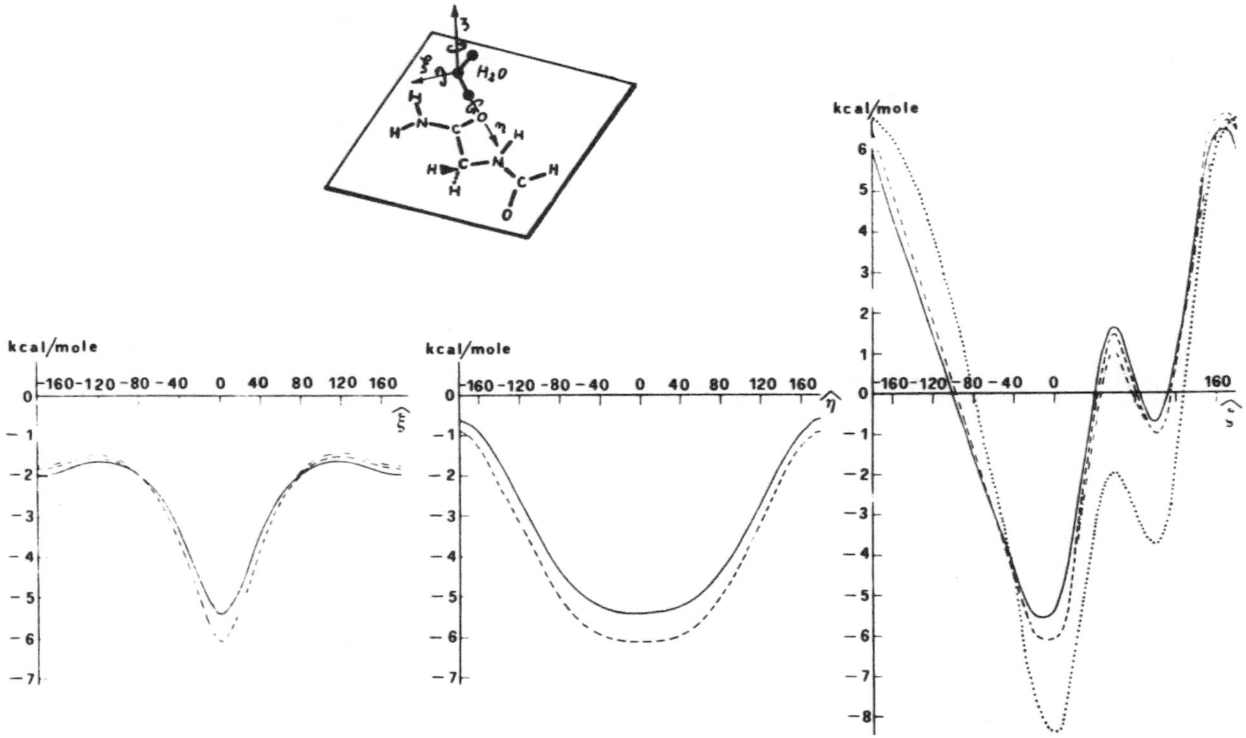

Fig. 26. Electrostatic calculations of the conformational curves for a 2-formylamino acetamide - water complex: ——— from the SCF wavefunction, ––––– from the LCAO expression of transferable group contributions, –·–·–·– from the point charge approximation of the transferable group contributions, ····· from the transferable gross atomic charge.

atomic charges for the other portion of the molecules. This last curve
shows again the inferiority of the transferable atomic charge models.

The approximations here outlined allow a noticeable enlargement of
the field of chemical systems where an application of electrostatic ap-
proachs relying on V are possible. To give an idea of the reduction of
the computation times, the calculations of curve b of figure 26 are
1.000 times shorter than the corresponding calculations on the supermolecule, and this ratio becomes even more favorable when one passes to
larger systems.

As we have seen above there is a wide range of molecular interac-
tive phenomena where such methods can find an application. Even the simplest non covalent interactions are of paramount importance in a wide
variety of situations, like the enzyme-substrate interactions in the basic reactions of the life cycle, the interactions between drugs and organic receptors, the interactions between organic catalysts and substra-
tes.

I would like to conclude this exposition by stressing that electrostatic calculations based on V cannot give, even in the most favourable
cases, anything more than partial answers to such problems which are
quite complex and which require, for their solution, the strict cooperation of researchers expert in different disciplines. An interesting
example of cooperation between molecular pharmacologists and theoretical
chemists can be found in a recent paper by Daudel et al. |48| concerning
a series of disubstituted 1,4-tetrahydro-oxazines: a positive feedback
between electrostatic calculations and experimental measurements per-
mitted more light to the shed on the pharmaceutical properties of this
interesting family of compounds.

REFERENCES

1. Thrular, D.G. and Van Catledge, F.A.: 1976, J. Chem. Phys. 65, pp.
   5536-5538.
2. Ghio, C., Scrocco, E., and Tomasi, J.: 1976, Jerusalem Symposia in
   Quantum Chemistry and Biochemistry, Vol. VIII "Environmental Effects
   on Molecular Structure and Properties", B.Pullman Ed., Reidel Pub.Co.,
   Dordrecht, Holland pp.329-342.
 - Julg, A.B. and Letoquart, D.: 1976, Phil. Mag. 33, pp.721-731.
3. Carozzo, L., Corongiu, G., Petrongolo, C., and Clementi, E.: 1978, J.
   Chem. Phys. 68, pp.787-793.
 - Mc Creery, Christoffersen, R.E., and Hall, G.C.: 1976, J. Amer. Chem.
   Soc. 98, pp.7191-7197 and 7198-7202.
4. Dreyfus, M. and Pullman, A.: 1970, Theor. Chim. Acta 19, pp.20-37.
5. Morokuma, K.: 1971, J. Chem. Phys. 55, 1236-1244.
6. Alagona, G., Scrocco, E., and Tomasi, J.: 1975, J. Amer. Chem. Soc.
   97, pp.6976-6983.
7. Umeyama, H. and Morokuma, K.: 1977, J. Amer. Chem. Soc. 99, pp.1316-
   1332.
8. Scrocco, E. and Tomasi, J.: 1978, Adv. Quantum Chem. 11, pp.116-193.
9. Kollman, P.: 1977, J. Amer. Chem. Soc. 99, pp.4875-4894.

10. Alagona, G., Pullman, A., Scrocco, E., and Tomasi, J.: 1973, Int. J. Peptide Protein Res. 5, pp.251-259.
11. Umeyama, H., Morokuma, K., and Yamabe, S.: 1977, J. Amer. Chem. Soc. 99, 330-343.
12. Alagona, G., Cimiraglia, R., Scrocco, E., and Tomasi, J.: 1972, Theor. Chim. Acta 25, pp.103-119.
13. Kollman, P., Mc Kelvey, J., Johansson, A., and Rothenberg, S.: 1975, J. Amer. Chem. Soc. 97, pp.955-965.
14. Fukui, K., Yonezawa, T., and Shingu, H.: 1952, J. Chem. Phys. 20, pp. 722-725.
15. Hudson, R.F. and Klopman, G.: 1967, Tetrahedron Lett. pp.1103-1108.
16. Salem, J.: 1968, J. Amer. Chem. Soc. 90, pp.543-552 and 553-566.
17. Fukui, K., Yonezawa, T., and Nagata, C.: 1957, J. Chem. Phys. 26, pp. 831-841.
18. Fukui, K., Yonezawa, T., Nagata, C., and Shingu, H.: 1954, J. Chem. Phys. 22, pp.1433-1442.
19. Umeyama, H. and Morokuma, K.: 1976, J. Amer. Chem. Soc. 98, pp.4400-4404.
20. Bonaccorsi, R., Scrocco, E., and Tomasi, J.: 1976, Theor. Chim. Acta 43, pp.63-73.
21. Ghio, C. and Tomasi, J.: 1973, Theor. Chim. Acta 30, pp.151-158.
22. Pullman, A.: 1973, Chem. Phys. Lett. 20, pp.29-32.
23. Bonaccorsi, R., Scrocco, E., and Tomasi, J.: 1976, J. Amer. Chem. Soc. 98, pp.4049-4054.
24. Pullman, A. and Brochen, P.: 1975, Chem. Phys. Lett. 34, pp.7-10.
25. Schuster, P., Jakubetz, W., and Marius, W.: 1975, Topics in Curr. Chem. 60, pp.1-107.
26. Cremaschi, P., Gamba, A., Morosi, G., Oliva, C., and Simonetta, M.: 1975, J. Chem. Soc. Faraday Trans. II pp.1829-1836.
27. Cremaschi, P., Gamba, A., and Simonetta, M.: 1975, Theor. Chim. Acta 40, pp.303-312.
28. Pasimeni, L., Corvaja, C., and Ghio, C.: 1978, Chem. Phys. 31, pp.31-37.
29. Politzer, P., Donnelly, R.A., Daiker, C.K.: 1973, J. Chem. Soc. Chem. Comm. pp.617-618.
30. Politzer, P. and Weinstein, H.: 1975, Tetrahedron 31, pp.915-923.
31. Bertran, J., Silla, E., Carbò, R., and Martin, M.: 1975, Chem. Phys. Lett. 31, pp.267-270.
32. Alagona, G., Scrocco, E., Silla, E., and Tomasi, J.: 1977, Theor. Chim. Acta 45, pp.127-136.
33. Alagona, G., Scrocco, E., and Tomasi, J.: "VIII Colloquio dei Chimici Teorici di Espressione Latina", Salamanca 1977.
34. Bonaccorsi, R., Scrocco, E., and Tomasi, J.: "VIII Colloquio dei Chimici Teorici di Espressione Latina", Salamanca 1977.
35. Dedieu, A. and Veillard, A.: 1972, J. Amer. Chem. Soc. 94, pp.6730-6738.
  - Bader, F.R.W., Duke, A.J., and Messer, R.R.: 1973, J. Amer. Chem. Soc. 95, pp.7715-7721.
  - Keil, F. and Ahlrichs, R.: 1976, J. Amer. Chem. Soc. 98, pp.4787-4793.

36. Leroy, G. and Sana, M.: 1975, Tetrahedron 31, pp.2091-2097.
   - Leroy, G. and Sana, M.: 1976, Tetrahedron 32, pp.709-717 and 1379-1382.
   - Leroy, G., Nguyen, N.T., and Sana, M.: 1976, Tetrahedron 32, pp. 1529-1536.
37. Daudel, R., Le Rouzo, H., Cimiraglia, R., and Tomasi, J.: 1978, Int. J. Quant. Chem. 13, pp.537-552.
38. Morokuma, K. and Iwata, S.: 1972, Chem. Phys. Lett. 16, pp.192-197.
39. Iwata, S. and Morokuma, K.: 1973, J. Amer. Chem. Soc. 95, pp.7563-7575.
   - Iwata, S. and Morokuma, K.: 1975, J. Amer. Chem. Soc. 97, pp.966-970.
40. Chow, Y.L.: 1973, Acc. Chem. Res. 6, pp.354-360.
41. Cimiraglia, R. and Tomasi, J.: 1977, J. Amer. Chem. Soc. 99, pp. 1135-1141.
42. Ghio, C., Scrocco, E., and Tomasi, J.: 1977, VIII Colloquio dei Chimici Teorici di Espressione Latina, Salamanca.
43. Bonaccorsi, R., Scrocco, E., and Tomasi, J.: 1974, Jerusalem Symp. on Quantum Chem. and Biochem., vol.VI "Chemical and Biochemical Reactivity", E.D.Bergmann and B.Pullman Eds, pp.387-395.
44. Giessner-Prettre, C. and Pullman, A.: 1972, Theor. Chim. Acta 25, pp.83-88.
45. Petrongòlo, C. and Tomasi, J.: 1973, Chem. Phys. Lett. 20, pp.201-206.
46. Bonaccorsi, R., Scrocco, E., and Tomasi, J.: 1976, J. Amer. Chem. Soc. 98, pp.4049-4054.
   - Bonaccorsi, R., Scrocco, E., and Tomasi, J.: 1977, J. Amer. Chem. Soc. 99, pp.4546-4554.
47. Hayes, H.D. and Kollman, P.: 1976, J. Amer. Chem. Soc. 98, pp.3335-3345 and 7811-7816.
48. Daudel, R., Esnault, L., Labrid, C., Busch, N., Moleyre, J. and Lambert, J.: 1976, Eur. J. Med. Chem. 11, pp.443-450.

NUCLEOPHILIC AFFINITY OF CONJUGATED HETEROCYCLES IN
PROTONATION, ALKYLATION AND CATION BINDING.

A. PULLMAN
Institut de Biologie Physico-Chimique

Laboratoire de Biochimie Théorique, associé au C.N.R.S.
13, rue Pierre et Marie Curie - 75005 Paris .

I. INTRODUCTION

The problem of the relative nucleophilicity of the various
sites in large heterocyclic molecules has always represented a difficult challenge for quantum chemistry. Although in the early times, the
classical indices of molecular structure were quite successful in accounting for the various kinds of reactivity of conjugated hydrocarbons, their utilization for heterocycles proved less straightforward.
The activity of these molecules towards electrophilic reagents is an
example where considerable energy has been spent on defining appropriate indices. It was soon recognized that, despite an apparent logical basis, correlations between the electronic charge of an atom and
its basicity (or more generally its nucleophilic reactivity) were very
limited. True already for the rating of molecules containing a single
heteroatom (quinolein with respect to isoquinolein, for instance), the
situation still worsens when a number of the same heteroatoms are present in the same compound, where the problem is then to discriminate
between their relative affinities for electrophilic attacks as for
instance (Figure 1) is the case for nitrogens N_1, N_3, N_7 of adenine,
N_3, N_7 of guanine, O_2 and O_4 of thymine or uracil; still more difficult is the discrimination between two sites of different nature in
the same molecule, for instance between a nitrogen in a ring, a nitrogen of an extracyclic amino group or a carbonyl oxygen.

When ab initio SCF computations on large molecules became
possible, it was hoped that the resulting charges, not suffering from
the incertainties of their semi-empirical predecessors, would proove
better indices of such local reactivities. However, (despite a recur-
rent belief) this is not the case : thus for instance, if one takes
the net electronic charges (from an ab initio SCF computation of reasonable accuracy (1)) as a measure of the basic character of a nitrogen, this would lead to predict the order $N_3 > N_7$ in guanine (negative
charges : 0.361, 0.247) whereas the reverse is known to be true experimentally. In the same molecule the charge of the oxygen (0.369) would

R. Daudel, A. Pullman, L. Salem, and A. Veillard (eds.), Quantum Theory of Chemical Reactions, Volume I, 229-241.
Copyright © 1979 by D. Reidel Publishing Company.

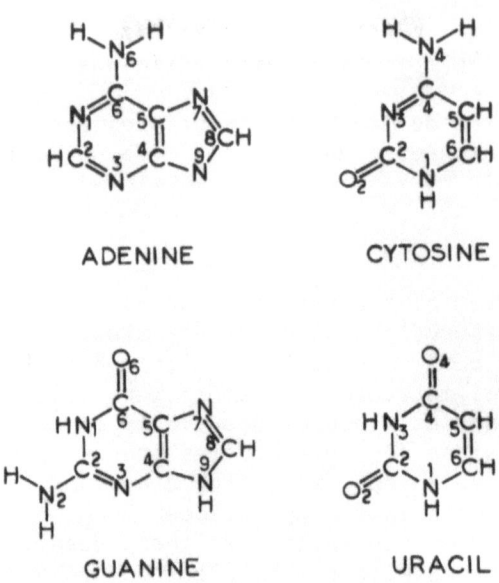

ADENINE   CYTOSINE

GUANINE   URACIL

Figure 1. The structure of the bases of the nucleic acids. (Thymine is
         5-methyluracil).

suggest erroneously that this atom should be more basic than N_7. Similarly, the negative charges of the amino nitrogen (0.614) or the imino
nitrogens (0.466 and 0.439) appreciably larger than those of N_3, N_7
and O_6 would suggest them to be the most basic centers, in contradiction with their known very small or negligible proton affinities.

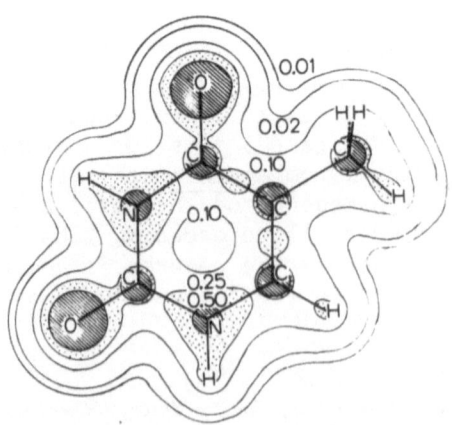

Figure 2. Electron density contours (e/a_o^3) in thymine. (Reproduced
          from reference (2), by permission.)

Another hope, placed in the utilization of more precise characteristics of the electron distribution such as the electron density maps turned out to be equally disappointing : see for instance (Fig. 2), the very similar aspect of the shape and density of the electron distribution (2) around the two oxygens of thymine, which prohibits a distinction between these nucleophilicities.

II. THE MOLECULAR ELECTROSTATIC POTENTIALS

A promising breakthrough towards the definition of an appropriate index for nucleophilicity came with the recognition (3) of the possibility of computing the molecular electrostatic potential created at all points in space by the set of nuclear charges and the electron distribution of a molecule. For a given wave function with the corresponding electron distribution $\rho(i)$ and a set of nuclei of charge Z_α, this potential at a given point P is :

$$(1) \qquad V(P) = \sum_\alpha \frac{Z_\alpha}{r_{\alpha P}} - \int \frac{\rho(i)}{r_{Pi}} \, d\tau_i$$

The molecular potential has two advantages over the classical indices of molecular structure (charges, free valence, etc...) : on the one hand it is a direct outcome of the wave function (through ρ) with no arbitrary partitionning (as opposed to the Mulliken (4) or other populations); on the other hand it is an expression of the global molecular reality which is clearly related to what a reagent "feels" upon approaching : the energy of interaction of the molecule, supposed unperturbed, with a point positive charge q placed where the potential is V, is equal to qV. If the charge is a proton the interaction energy is equal to V.

In reality, the true energy of interaction is not strictly qV because the approaching charge polarizes the molecule, but this modification can be considered as a perturbation. In fact, in terms of perturbation theory, it is easily seen that the energy qV is the first-order term of the interaction energy, the polarization term intervening at second order only, together with a "charge-transfer" (CT) term which corresponds to the transfer of electrons towards the approaching charge resulting in the final formation of a chemical bond. Hence, the total energy of, say, protonation can be written as a sum of three terms (5) :

$$(2) \qquad \Delta E = V + \Delta E_{POL} + \Delta E_{CT}$$

where V is the value of the potential.

This analysis shows that in spite of its being a first order quantity, the potential is an index of molecular structure which has a logical relation to the reactivity of the molecule towards an approaching charged species. In terms of the usual vocabulary of quantum chemistry it should be classified as a "static" index, characterizing the unperturbed molecular structure, whereas the polarization and charge transfer energies are "dynamic indices". It presents however one fundamental advantage with respect to the classical static indices in that it represents the expression, at a given point in space, of the effect of the <u>whole</u> molecular structure, in strong contrast to all the classical indices which are characteristics of local properties. Now, it was shown in 1958 (6) that the protonation of large multicenter heterocycles could not be understood without taking into account the interaction of the proton with the entire electron distribution of the molecule. This finding takes its full meaning in the light of the notion of molecular potential.

The question which we wish to investigate particularly in this paper concerns the evaluation of the extent to which this new notion of molecular electrostatic potential can be used as an index of nucleophilicity. It is related in particular to a series of very fundamental theoretical and experimental studies carried out in a number of laboratories on the nucleophilicity of the heteroaromatic rings of the purine and pyrimidine bases of the nucleic acids,pertaining to their reactivity towards carcinogenic, mutagenic, chemotherapeutical, etc, electrophiles. For this sake we shall study here the reactivity of these bases towards protonation, cation (Na^+) binding and two types of alkylation, by a methyl and an ethyl cation. All these reactions result in the fixation of the cation upon the attacked substrate. Our goal is to put into evidence the analogies and /or differences, if any, in the mechanisms and products of these transformations.

III. PROTONATION

One of the first successes of the utilization of the molecular potential was the determination of the protonation sites in the nucleic bases (7 - 8 - 9).

The best way to use potential is to draw isoenergy contours in the space surrounding the molecule. An example, the case of guanine, is given in Figure 3 a,b,c. The diagrams present attractive and repulsive regions for a point positive charge and well-defined potential wells converging towards negative minima. In guanine, the most important minima are located in the molecular plane in the neighbourhood of N_7, O_6 and N_3,the largest negative value occuring in the neighbourhood of N_7. Noticeable is the fact that the largest negative domain encompasses both the O_6 and N_7 regions with a distinct minimum close to O_6. It is quite typical of this kind of molecules (planar conjugated heterocycles) that the most attractive regions are located in the molecular plane, although the out-of-plane regions do present also attractive zones sometimes converging towards a minimum.

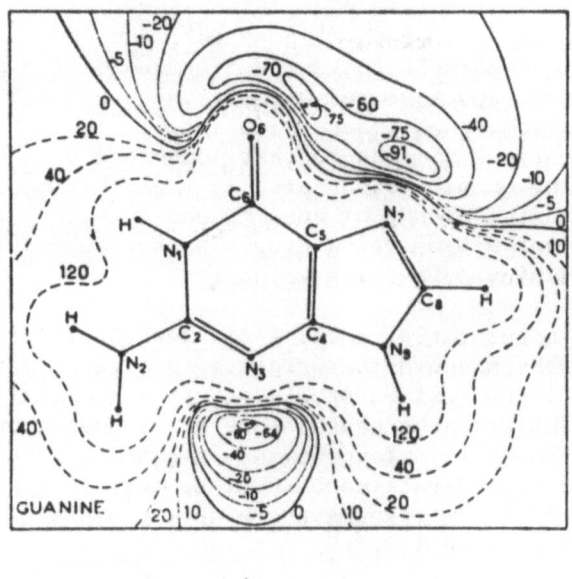

(a)

(b)   (c)

Figure 3. Isopotential maps a) in the plane of guanine; b) in a perpendicular plane containing C_8H; c) in a perpendicular plane containe C_2N_2. (Reproduced from ref. (9) by permission).

However the corresponding values, are as a rule, much smaller than the
in-plane ones : in the case of guanine, a very small negative minimum
of this kind is seen above C_8 (Figure 3b). Another position of interest is shown in Figure 3c above the nitrogen of the amino group : this
region is in fact repulsive for a positive charge but the shape of the
isoenergy lines converges towards a point of "least repulsion" of +2
kcal/mole. Situations similar to that found above C_8 and above the
amino nitrogen of guanine occur also in the other bases : for instance
above C_5 of cytosine or NH_2 of adenine and cytosine, the depths of
these out-of-plane minima being always appreciably smaller than those
of the in-plane minima in these molecules.

From a quantitative point of view the location of the dee-
pest minima has been shown to correlate in a very satisfactory way with
the position of the most favorable sites of protonation in these molecules which indicate the preeminence of N_1 and N_3 over N_7 in adenine
with a strong reversal in favor of N_7 in guanine, the dominance of N_3
in cytosine, of O_4 in uracil and thymine, a general dominance of the
ring nitrogens over the oxygens and over the amino or imino nitrogens
in all these bases (see 1,7,8,9).

Whenever the complete protonation energies were computed in
these molecules (10-11) they gave the same indications as to the most
basic site as the molecular potentials. The reason of this agreement
is that, in this kind of molecules, the supplementary terms which
enter in equation (2), although by no means negligible, either vary
in the same direction as the molecular potentials, or present variations which are not strong enough to conteract the tendancy set out
by the electrostatic term (see (5)).

Note that this situation is valid for the type of molecules
considered here (conjugated heterocyclic molecules or analogs) and
need not be universal. There are cases where the proton affinities and
the molecular potentials vary in opposition and this is quite understandable on the basis of equation (2).

IV. BINDING OF Na^+

Na^+ is one of the simplest metal cations which are of interest from the point of view of their possible affinity for the components of the nucleic acids. The question which arises naturally after
the observations concerning the proton affinity of the bases is whether a cation such as Na^+, will show the same preference as a proton
in the choice of its site of attachement. In other words, does the interaction with Na^+ follow the indications given by the molecular potential as to the most nucleophilic position on the bases.

The Na^+-affinity for the nucleic bases has been studied by
the supermolecule approach at the SCF level (12). The results are best

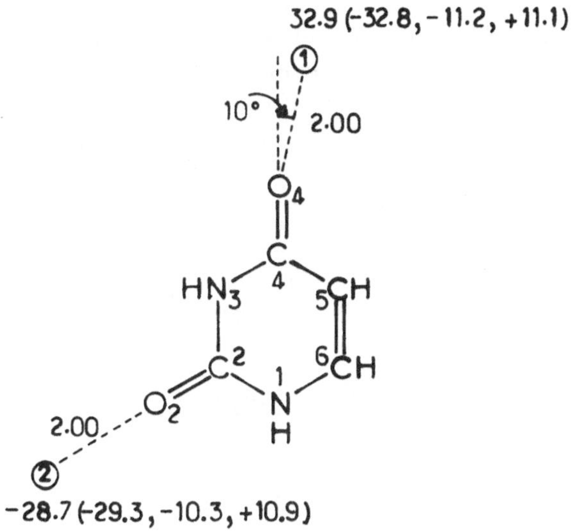
32.9 (-32.8, -11.2, +11.1)
1
10°
2.00
O_4
C
4
HN_3
5CH
2
6CH
C_2
1
N
H
2.00
O_2
2
-28.7 (-29.3, -10.3, +10.9)

Figure 4. Affinity of uracil for Na^+. Equilibrium positions and energy components.

illustrated on the examples of uracil and cytosine. In uracil two favorable equilibrium positions are found for the fixation of the sodium cation (Figure 4), one in the neighbourhood of O_4 the other in the neighbourhood of O_2, the binding to O_4 being favored. Thus in uracil, the situation relative to Na^+-binding is exactly that which would be predicted by the order of the minima in the map of the molecular potential (1). The reason for this situation appears upon performing a decomposition of the binding energy into its components. In a fashion similar to that leading to equation (2) the binding energy of a molecule with another entity can be expressed as a sum (13)

$$(3) \qquad \Delta E = \Delta E_C + \Delta E_{DEL} + \Delta E_{EX}$$

where ΔE_C is the pure electrostatic interaction of the unperturbed partners, ΔE_{DEL} the sum of the polarization and charge-transfer contribution and ΔE_{EX} a repulsive term which arises from the overlap of the electron distributions of the two interacting species. This term does not exist when the approaching ligand is a proton which carries no electrons.

The values of these different components inside the binding

energies of Na^+ to uracil show the predominance of the electrostatic
term. Moreover, at the equilibrium distance with respect to O_4 and O_2,
the repulsive exchange and the attractive delocalization energies essentially cancel each other, so that the total energy is practically
equal to the Coulomb term. It is thus understandable that the map of
electrostatic attraction for a point charge predicts with a very good
approximation the location of the best positions and the order of the
binding energies. A very similar situation was encountered for the
binding of Na^+ to adenine (12).

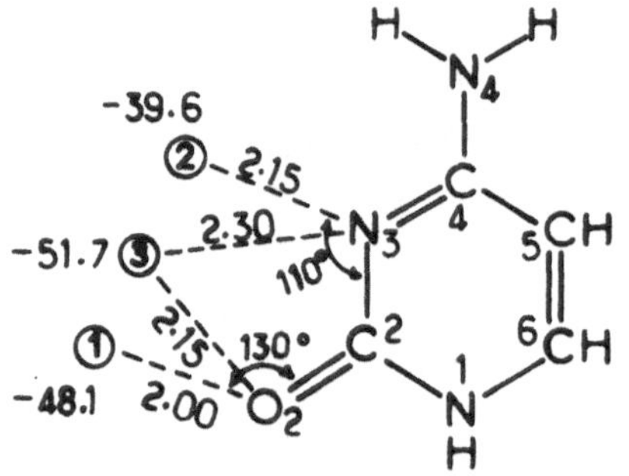

Figure 5. Affinity of cytosine for Na^+. Equilibrium distances and ener-
          gies.

        Going over to guanine and cytosine reveals however an inte-
resting differing phenomenon : this situation is illustrated on the
example of cytosine (Figure 5 and Table I). Let us restrict first the
approach of Na^+ towards N_3 or O_2 along the directions leading to the
minima of the molecular potential map : it is seen that in this mode
of approach, Na^+ finds in each case an equilibrium position with a binding energy appreciably larger for site 1 (oxygen binding) than for
site 2 (nitrogen binding), with optimal distances of 2 and 2.15 Å respectively. This shows that, in cytosine, the carbonyl oxygen exerts
an overall stronger attraction than the nitrogen towards the cation,
contrary to the indications that would be given by considerations based on the molecular potential. Again, the explanation of the situation is found upon the examination of the components of the binding
energy : it is seen that for Na^+ situated at the same distance of N_3
and O_2, the pure coulomb component is nearly equal for the two positions and the same is true for the delocalization energy, hence for
the global attractive part of the energy; but on the other hand, the

repulsive exchange term is appreciably larger for N_3 than for O_2. This explains that the optimal distance of Na^+ to N_3 is larger than that of Na^+ to O_2 and that the final binding energy is larger for an attachment to O_2 than for an attachment to N_3.

Hence in cytosine the relation between the cation-binding preference and the molecular potential map is different from that observed in uracil because in cytosine the coulomb component of the binding energy is no longer decisive in the <u>discrimination</u> between the two kinds of atoms (although it still is numerically dominating the total energy). One may say that in this case, the discrimination between the oxygen and nitrogen is done by a criterion of "least repulsion". A similar situation occurs in guanine as far as the intrinsic preference of Na^+ for O_6 and N_7 is concerned.

If one now lets the sodium cation free to move to the most favorable site it will end up in a bridge position between the oxygen and the nitrogen (Figure 5). This position is closer to O_2 than to N_3, in terms of the preceeding analysis. The possibility to adopt such a bridge position is due to the particular dimension of Na^+.

TABLE I
Binding energies and their components (kcal/mole) for Na^+-uracil (see Figure 4 for geometry).

Site	distance	ΔE	ΔE_C	ΔE_{DEL}	ΔE_{EX}
1	O_2Na^+ = 2.0	-48.09	-50.12	-10.90	12.93
	O_2Na^+ = 2.15	-47.90	-44.35	-9.13	5.58
2	N_3Na^+ = 2.0	-35.95	-51.62	-10.48	26.15
	N_3Na^+ = 2.15	-39.59	-43.60	-8.74	12.75
3	N_3Na^+ = 2.30 O_2Na^+ = 2.15	-51.73	-51.08	-12.31	11.66

In conclusion it appears that in the choice between sites of the same nature (oxygens of uracil, nitrogens of adenine) the preference of Na^+ is, like the preference of H^+, essentially dictated by the coulomb attraction. Between two sites of different nature, like the O and the N, the intrinsic preference of Na^+ is reversed with respect to that of a proton and this result is due to the exchange repulsion, which is fundamentally larger for nitrogen than for oxygen. This is

understandable if one remembers that the valence shell orbitals of a
nitrogen atom extend farther in space than those of an oxygen atom,
thus giving rise to a larger overlap with those of the attacking ion
at a given distance, hence to a larger repulsion.

V. ALKYLATION

Alkylating agents are an important class of electrophiles,
defined for the sake of our present discussion as reagents RX, which
react with a molecule M in such a way as to form a product MR$^+$, R
being an alkyl group. Similar arylating agents exist where R is an
aromatic residue.

TABLE II
Binding energies and their components (kcal/mole) of H^+, CH_3^+, $C_2H_5^+$
to N_3 and O_2 of cytosine at equilibrium distance d(Å). ΔE relative
to undeformed ion, $\Delta E'$ relative to ion in its final shape.

		d	ΔE	ΔE_C	ΔE_{DEL}	ΔE_{EX}	$\Delta E'$
H^+	N_3		−293.3	−105.4	−187.9		
	O_2		−291.5	−85.6	−105.9		
CH_3^+	N_3	1.49	−170.7	−136.6	−199.3	134.3	−201.7
	O_2	1.46	−171.0	−105.5	−201.9	105.4	−202.0
$C_2H_5^+$	N_3	1.50(a)	−124.0				
		(b)	−134.8	−136.6	−193.2	178.8	−151.0
	O_2	1.48(a)	−140.2	−104.7	−187.8	136.0	−156.4
		(b)	−140.5				

(a) C'C" bond of ion in the cytosine plane
(b) C'C" bond of ion rotated 90° about NC' or OC' bond.

The reactions of the components of the nucleic acids with
alkylating agents have attracted particular attention because of their
involvment in mutagenesis and carcinogenesis. In particular, the reactions of the purine and pyrimidine bases with various alkylating
agents have been and are more and more extensively studied. One of the
recent developments in this field is the progressive demonstration

that different classes of alkylating agents differ in their affinity
for tarjet sites on the bases : thus for instance, in guanine, while
N_7 is the preferred site of methylation under neutral conditions,
O_6 becomes the preferred site of ethylation with certain ethylating
agents such as ethylnitrosourea (but not with others such as ethyl-
methanesulfonate) (14). More recently it was shown that while cyti-
dine reacts with a variety of methylating agents at N_3 (where it also
protonates) the major product of ethylation by ethylnitrosourea is
O_2-ethylcytidine (15).

Without entering here into details about the mechanism of
action specific for the different types of alkylating agents, it is
interesting, in view of these differences, to explore from the theoretical point of view what is the intrinsic preference of a methyl and
of an ethyl cation with respect to attachement on a nitrogen or an
oxygen atom of a nucleic base. This was done on the example of cytosine
(10). An interesting evolution of the situation when going from H^+ to
CH_3^+ and $C_2H_5^+$ is seen in comparing the results of Table II. Thus protonation is found to favor the nitrogen over the oxygen position, essentially owing to the coulomb attractive term. Going over to CH_3^+,
one observes that the binding energies at equilibrium (with respect to
the reactants infinitely separated) become practically identical for
the methylation of N_3 and O_2. The components of the binding energy for
CH_3^+ as for H^+ reveal that the pure electrostatic attraction is larger
towards the nitrogen atom than towards the oxygen, that the delocalization term is very similar for the two positions, but that the exchange repulsion is appreciably larger for an approach towards N_3 than
towards O_2 (in spite of the larger distance) so that a practical cancellation of tendancies occurs, making the binding energy indiffe-
rent to the position of attack. Finally for the ethyl cation the preference observed in the final equilibrium values is now in favor of
the oxygen. The factors involved appear again in the calculation of
the components of the binding energy (Table II) : at equilibrium, the
pure coulomb attraction would favor the nitrogen, the delocalization
adding somewhat to this trend, but the repulsion strongly counteracts
this tendency in such a way as to make the total energy more favorable
for fixation on oxygen. The importance of the exchange repulsion in the
phenomenon is further evidenced by the fact that the rotation of the
C'C" bond out of the cytosine plane improves the energy for addition
at the N position, obviously by decreasing the overall overlap, this
phenomenon being absent for rotation at the oxygen position.

Hence, it appears that the intrinsic preference of H^+, CH_3^+
and $C_2H_5^+$ towards the two essential possible binding sites in cytosine,
N_3 and O_2, undergo a continuous modification with the increase in size
and complexity of the cation, the phenomenon being due to the increasing effect of the exchange repulsion counteracting the intrinsically
larger attraction for the nitrogen position. As was observed in the
analysis of the reasons of the intrinsic preference of Na^+ for oxygen
over nitrogen, the exchange repulsion is the crucial element in the

evolution of the reaction from H^+ to $C_2H_5^+$ fixation. A computational
control of the fact that this factor is essentially a characteristic
of the nitrogen and oxygen atoms (and not a result of the proximity
of the close NH_2 group to the nitrogen atom in cytosine) is shown in
Figure 6 where the exchange repulsion part of the interaction energy
between a CH_3^+ and the nitrogen atom of pyridine is plotted against
the N-C distance of approach in comparison to the exchange repulsion
part in the interaction energy of CH_3^+ with formaldehyde plotted
against the O-C distance : at a given distance the exchange term is
always larger for pyridine than for formaldehyde.

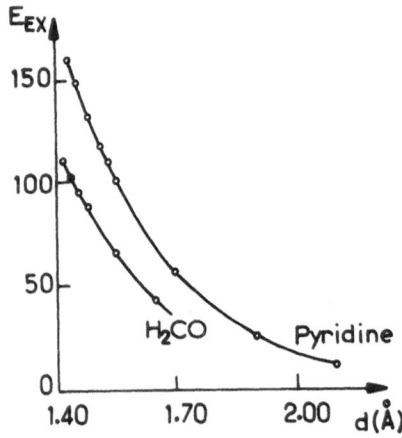

Figure 6. Distance dependence of the exchange repulsion in the binding
energy of CH_3^+ approaching in-plane towards the nitrogen of
pyridine and axially towards the oxygen of formaldehyde.

VI CONCLUSION

        The present analysis shows the complexity of the factors
involved in the interaction of different types of electrophilic reactants (protons, metal cations, alkylating species) with conjugated
heterocycles possessing a number of different potential nucleophilic
sites. It substantiates the occurrence of different types of products.
It shows that the electrostatic molecular potential of the substrate,
while providing useful information on its overall nucleophilicity need
not correlate with the nature of the products susceptible to be obtained with a variety of attacking electrophiles, due to the significant
role of factors other than the purely electrostatic ones in the development and termination of the reactions. It seems also to indicate
that the computation of the binding energy at equilibrium and the analysis of its components provide a correct prediction of the nature of
the products to be expected and an explanation for their formation.

REFERENCES

1 Pullman, A., in Mécanismes d'altération et de réparation du DNA. Relations avec la mutagénèse chimique, Colloque du CNRS n°356 à la mémoire de P. Daudel (1976)pp. 102-113.

2 Pullman, A., Dreyfus, M. and Mély, B. : 1970, Theor. Chim. Acta, 16, pp. 85-88.

3 Bonaccorsi, R., Scrocco, E. and Tomasi, J. : 1970, J. Chem. Phys. 52, pp. 5270-5283.

4 Mulliken, R.S.: 1955, J. Chem. Phys., 23, pp. 1833-1842.

5 Pullman, A. : 1977, Chem. Phys. Letters, 20, pp. 29-32.

6 Nakajima, T. and Pullman, A. : 1958, J. Chim. Phys. 55, pp. 793-801.

7 Bonaccorsi, R., Pullman, A., Scrocco, E. and Tomasi, J. : 1972, Theor. Chim. Acta, 24, pp. 51-60.

8 Giessner-Prettre, C. and Pullman, A. : 1971, C.R. Acad. Sci. Paris, C272, pp. 750-753.

9 Bonaccorsi, R., Scrocco, E., Tomasi, J. and Pullman, A. : 1975, Theor. Chim. Acta, 36, pp. 339-344.

10 Pullman, A. and Armbruster, A-M., 1977 : C.R. Acad. Sci. Paris, D284, pp. 231-234.

11 Pullman, A. and Armbruster, A-M., 1977 : Theor. Chim. Acta, 45, pp. 249-256.

12 Perahia, D., Pullman, A. and Pullman, B. : 1977, 43, pp. 207-214.

13 Dreyfus, M. and Pullman, A. : 1970, Theor. Chim. Acta, 19, pp. 20-37.

14 Singer, B. and Fraenkel-Conrat, H. : 1975, Biochem. 14, pp. 772-782.

15 Singer, B. : 1976, FEBS Letters, 63, pp. 85-88.

INDEX OF SUBJECTS